A Level Physics for OCR Year 2 Revision Guide

Gurinder Chadha

OXFORD
UNIVERSITY PRESS

Great Clarendon Street, Oxford, OX2 6DP, United Kingdom

Oxford University Press is a department of the University of Oxford. It furthers the University's objective of excellence in research, scholarship, and education by publishing worldwide. Oxford is a registered trade mark of Oxford University Press in the UK and in certain other countries

© Gurinder Chadha 2017

The moral rights of the authors have been asserted

First published in 2017

All rights reserved. No part of this publication may be reproduced, stored in a retrieval system, or transmitted, in any form or by any means, without the prior permission in writing of Oxford University Press, or as expressly permitted by law, by licence or under terms agreed with the appropriate reprographics rights organization. Enquiries concerning reproduction outside the scope of the above should be sent to the Rights Department, Oxford University Press, at the address above.

You must not circulate this work in any other form and you must impose this same condition on any acquirer

British Library Cataloguing in Publication Data
Data available

978 0 19 835778 0

10 9 8 7 6 5 4 3 2 1

Paper used in the production of this book is a natural, recyclable product made from wood grown in sustainable forests. The manufacturing process conforms to the environmental regulations of the country of origin.

Printed in Great Britain

COVER: Bizroug / Shutterstock

Artwork by Q2A Media

Contributors

Geneva Beck started beading on doll costumes and now loves designing jewelry. Contact her via email at gbtimsbelle@aol.com.

Hannah Benninger learned about cubic right-angle weave from an article about bead artist David Chatt in *The Art of Beadwork: Historic Inspiration, Contemporary Design* by Valerie Hector, and has designed many projects using this technique. Contact her via email at hannah.benninger@gmail.com.

When not managing an audit department, **Connie Blachut** enjoys the creative expression of designing and making jewelry at her home in Plymouth, Mich. Contact her via email at cblachut07@comcast.net.

April Bradley, originally from Alaska, now lives in Valley Forge, Pa., with her husband and children. Contact April via email at aprilbradley@comcast.net, or visit her website, aprilbradley.com.

Abby Cobb has been beading since 2002. She prefers to use inexpensive materials such as crystals and seed beads to make beautiful, affordable pieces. Contact Abby via email at ajcobb_10@yahoo.com.

Teri Dannenberg began beading with semiprecious stones because she was fascinated by their natural patterns and colors. Later, she discovered even more ways to play with color and texture using seed beads and crystals. Contact her via email at teridann@gmail.com.

Marcia DeCoster makes her living by, as well as derives much pleasure from, pursuing a beady life. She has been beading and teaching worldwide since 1999. Her book *Marcia DeCoster's Beaded Opulence* is part of the Beadweaving Master Class series. Visit her website, marciadecoster.com.

Seed bead weaver and beading instructor **Donna Pagano Denny** is always working with her hands. She's crocheted and tatted with beads, and seed beads are her passion. Her motto is "the smaller the beads the better." Donna enjoys sharing her craft with fellow beaders in Fayetteville, N.C., and sells her work at Lush Beads in her hometown of Lowell, Mass. Contact her via email at lacetatter@aol.com.

Phyllis Dintenfass designs, publishes, and teaches off-loom beadwork. Her finished jewelry has been exhibited in juried shows in the U.S. and abroad. Visit her website, phylart.com, or contact her via email at phylart@new.rr.com.

Julia Gerlach is Editor of *Bead&Button* magazine. Contact her at jgerlach@kalmbach.com.

Julie Glasser has been beading since 1980, when she inherited her grandmother's wire and seed beads. Now she focuses on beadweaving techniques and teaches beading classes at an art school in Atlanta, Ga. She is also an accomplished metalsmith, combining sterling silver and seed beads in a lot of her work. Visit Julie's website at julieglasser.com.

Mia Gofar is the author of several how-to jewelry books in Indonesia. Her first book was released in 2005, and she continues to write books today. Contact Mia via email at mia@miagofar.com, or visit her website, miagofar.com or miamoredesign.com.

Lisa Kan is a beader and lampwork artist who enjoys incorporating innovative components and seed beads in designing distinctively elegant beadwork with a vintage feel. She draws her inspiration from nature, history, ceramics, Japanese arts and aesthetics, and Victorian-era jewelry, and is the author of *Bead Romantique: Elegant Beadweaving Designs*. Contact Lisa through her website, lisakan.com.

Barbara Klann has been beading for more than 20 years, and is constantly amazed at what can be accomplished with a small pile of beads and a needle and thread. Her main materials of choice are seed beads with a few crystals or pearls sprinkled in here and there. Contact Barbara in care of Kalmbach Books.

Cathy Lampole of Newmarket, Ontario, Canada, enjoys the fine detail that can be achieved with bead weaving, especially with crystals. Besides designing jewelry, Cathy owns a bead shop, That Bead Lady. Visit her website, thatbeadlady.com, or email her at cathy@thatbeadlady.com.

Shelley Nybakke lives in Normal, Ill. She travels and teaches workshops around the country, and thinks a day without beads is hardly ever worth getting out of bed for. Contact Shelley via email at shelley@thebeadparlor.com or visit her website, shelleynybakke.com.

Pam O'Connor lives with her family in Naples, Italy, and travels frequently for inspiration and acquisition of beads and other embellishments. Contact her via email at erzulimojo@gmail.com or visit her online at erzulimojo.etsy.com.

Cindy Thomas Pankopf teaches beading in Southern California and at the Bead&Button Show, and is also a Master instructor for Art Clay World. She is the author of *BeadMaille* and *The Absolute Beginners Guide: Making Metal Clay Jewelry*. Contact her via email at info@cindypankopf.com, or visit her website, cindypankopf.com.

Michelle Skobel has been beading for most of her life and started designing jewelry three years ago. You can see more of her work, kits, and patterns on her website, michelleskobel.com, or email her at michelle@michelleskobel.com.

Deborah Staehle began beading in the mid-1990s as a member of the Bead Society of Hawaii, Oahu chapter. It was meant to be just a hobby. But after moving back to California in 2001 she began working and teaching full time in the bead business. You can find her six days a week at Bead Dreams in Stockton, where she teaches and sells beads. Contact her at Bead Dreams at (209) 464-2323 or via email at bead_demon@hotmail.com.

Email **Barb Switzer** at barb@beadswitzer.com or visit her website, beadswitzer.com.

Lisa Twede lives in Burbank, Calif. She loves being reminded of the joy she experiences making jewelry when she sees her friends wearing pieces she made for them. Visit her at lisabearsbeadingblog.blogspot.com, or follow "Awesome Beading Patterns" on Facebook.

Julie Walker currently lives in Dayton, Ohio, and shares a bead business, Viking Fiber and Beads (formerly The Bead Cage), with friend and fellow designer Suzie Jaisle. She loves creating new and exciting ideas for her beady customers. Julie is the proud mother of four daughters and nana to two. Visit her website, beadcage.net, or email her at beadcage@gmail.com.

Lesley Weiss is the author of *The Absolute Beginners Guide: Stitching Beaded Jewelry* and a former associate editor for *Bead&Button* magazine. She has been beading since 2003, and loves how stitching with beads engages her analytical, puzzle-solving skills and indulges her creative side at the same time. Visit her website, homemade-handmade.net.

Nancy Zellers is the author of the book *Bead Tube Jewelry*. To contact her or see more of her work, visit her website, nzbeads.com.

Pursue Your Passion

The Start of a Terrific Library!

The amazing fact that one stitch can lead to so many possibilities is the foundation of the *Stitch Workshop* series. The first book in the series, *Peyote Stitch*, illustrates this by delving deep into the popular stitch and then supplying over 25 of the best projects from *Bead&Button* magazine, so beginners and advanced bead artists can perfect their techniques.

64230 $17.95

Look for more books to come in the *Stitch Workshop* series!

Buy now from your favorite bead or craft shop!
Or at **www.KalmbachStore.com** or call **1-800-533-6644**
Monday – Friday, 8:30 a.m. – 4:30 p.m. CST. Outside the United States and Canada call 262-796-8776, ext. 661.

AS/A Level course structure

This book has been written to support students studying for OCR A Level Physics A. It covers Year 2 modules from the specification. These are shown in the contents list, which also shows you the page numbers for the main topics within each module.

FREE BOOKS
HATFIELD

AS exam

Year 1 content
1. Development of practical skills in physics
2. Foundations in physics
3. Forces and motion
4. Electrons, waves, and photons

Year 2 content
5. Newtonian world and astrophysics
6. Particles and medical physics

A level exam

A Level exams will cover content from Year 1 and Year 2 and will be at a higher demand. You will also carry out practical activities throughout your course.

Contents

How to use this book vi

Module 5 Newtonian world and astrophysics 1
14 Thermal physics 1
14.1 Temperature 2
14.2 Solids, liquids, and gases 2
14.3 Internal energy 4
14.4 Specific heat capacity 4
14.5 Specific latent heat 4
Practice questions 6

15 Ideal gases 7
15.1 The kinetic theory of gases 8
15.2 Gas laws 8
15.3 Root mean square speed 10
15.4 The Boltzmann constant 10
Practice questions 12

16 Circular motion 13
16.1 Angular velocity and the radian 14
16.2 Centripetal acceleration 14
16.3 Exploring centripetal forces 16
Practice questions 18

17 Oscillations 19
17.1 Oscillations and simple harmonic motion 20
17.2 Analysing simple harmonic motion 22
17.3 Simple harmonic motion and energy 22
17.4 Damping and driving 24
17.5 Resonance 24
Practice questions 26

18 Gravitational fields 27
18.1 Gravitational fields 28
18.2 Newton's law of gravitation 28
18.3 Gravitational field strength for a point mass 28
18.4 Kepler's laws 30
18.5 Satellites 30
18.6 Gravitational potential 32
18.7 Gravitational potential energy 32
Practice questions 34

19 Stars 35
19.1 Objects in the Universe 36
19.2 The life cycle of stars 36
19.3 The Hertzsprung-Russell diagram 36
19.4 Energy levels in atoms 38
19.5 Spectra 38
19.6 Analysing starlight 40
19.7 Stellar luminosity 40
Practice questions 42

20 Cosmology (the Big Bang) 43
20.1 Astronomical distances 44
20.2 The Doppler effect 44
20.3 Hubble's law 46
20.4 The Big Bang theory 46
20.5 Evolution of the Universe 46
Practice questions 48

Module 6 Particles and medical physics 49
21 Capacitance 49
21.1 Capacitors 50
21.2 Capacitors in circuits 50
21.3 Energy stored by capacitors 50
21.4 Discharging capacitors 52
21.5 Charging capacitors 52
21.6 Uses of capacitors 52
Practice questions 54

22 Electric fields 55
22.1 Electric fields 56
22.2 Coulomb's law 56
22.3 Uniform electric fields and capacitance 58
22.4 Charged particles in uniform electric fields 58
22.5 Electric potential and energy 60
Practice questions 62

23 Magnetic fields 63
23.1 Magnetic fields 64
23.2 Understanding magnetic fields 64
23.3 Charged particles in magnetic fields 66
23.4 Electromagnetic induction 68
23.5 Faraday's law and Lenz's law 68
23.6 Transformers 68
Practice questions 70

24 Particle physics 71
24.1 Alpha-particle scattering experiment 72
24.2 The nucleus 72
24.3 Antiparticles, hadrons, and leptons 74
24.4 Quarks 74
24.5 Beta decay 74
Practice questions 76

25 Radioactivity 77
25.1 Radioactivity 78
25.2 Nuclear decay equations 78
25.3 Half-life and activity 80
25.4 Radioactive decay calculations 80
25.5 Modelling radioactive decay 82
25.6 Radioactive dating 82
Practice questions 84

26 Nuclear physics 85
26.1 Einstein's mass-energy equation 86
26.2 Binding energy 86
26.3 Nuclear fission 88
26.4 Nuclear fusion 88
Practice questions 90

27 Medical imaging 91
27.1 X-rays 92
27.2 Interaction of X-rays with matter 92
27.3 CAT scans 94
27.4 The gamma camera 94
27.5 PET scans 94
27.6 Ultrasound 96
27.7 Acoustic impedance 96
27.8 Doppler imaging 96
Practice questions 98

A1 Physics quantities and units 99
A2 Recording results and straight lines 99
A3 Measurements and uncertainties 100
Physics A data sheet 102
Answers to practice questions 105
Answers to summary questions 110

How to use this book

> **Specification references**
> → At the beginning of each topic, there are specification references to allow you to monitor your progress.

Revision tips
Prompts to help you with your understanding and revision.

Synoptic link
These highlight the key areas where topics relate to each other. As you go through your course, knowing how to link different areas of physics together becomes increasingly important. Many exam questions, particularly at A Level, will require you to bring together your knowledge from different areas.

This book contains many different features. Each feature is designed to support and develop the skills you will need for your examinations, as well as foster and stimulate your interest in physics.

 Worked example
Step-by-step worked solutions.

Common misconception
Common misunderstandings clarified.

Maths skills
A focus on maths skills.

Model answers
Sample answers to exam-style questions.

Summary Questions

1. These are short questions at the end of each topic.

2. They test your understanding of the topic and allow you to apply the knowledge and skills you have acquired.

3. The questions are ramped in order of difficulty. Lower-demand questions have a paler background, with the higher-demand questions having a darker background. Try to attempt every question you can, to help you achieve your best in the exams.

Chapter 14 Practice questions

1. Two objects X and Y are in thermal equilibrium. Which statement is correct?
   A There is a net transfer of energy between X and Y.
   B X and Y have the same amount of thermal energy.
   C X and Y have the same temperature in kelvin.
   D X and Y have the same specific heat capacity. *(1 mark)*

2. What is specific latent heat of fusion measured in base units?
   A $m\,s^{-2}$
   B $m^2\,s^{-2}$
   C $m^2\,s^{-2}\,K^{-1}$
   D $kg\,m^2\,s^{-2}$ *(1 mark)*

3. Four solids A, B, C, and D are heated using the same heater. The solids have the same mass.
   A temperature–time graph is plotted for each solid on the same axes (see Figure 1).
   Which solid has the smallest value of specific heat capacity? *(1 mark)*

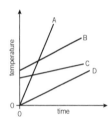
▲ Figure 1

4. A beaker with 120 g of water at 15 °C is placed inside a freezer. It takes 6.0 minutes for the temperature of the water to drop to 0 °C.
   specific heat capacity of water = $4200\,J\,kg^{-1}\,K^{-1}$
   specific latent heat of fusion of ice = $3.3 \times 10^5\,J\,kg^{-1}$
   a Calculate the average rate of energy loss from the water. *(3 marks)*
   b Calculate the time it would take for the water at 0 °C to turn into ice at 0 °C. Assume that the rate of energy transfer is the same as your answer to **a**. *(3 marks)*

5. Figure 2 shows a heater used to melt some crushed ice in a funnel.
   The power of the heater is 25 W. The temperature of the ice remains at 0 °C. The table below shows the variation with time t of the total mass m of the water from the melting ice.

t / s	0	30	60	90	120	150
m / g	0	2.7	5.5	8.2	10.9	14.0

   a Plot a graph of m against t and draw a line of best fit through the data points. *(2 marks)*
   b Use the graph to determine the gradient of the straight line. *(1 mark)*
   c Use your answer to **b** to calculate the specific latent heat of fusion of ice. *(3 marks)*
   d Explain why your value the specific latent heat of fusion of ice is smaller than the accepted value of $3.3 \times 10^5\,J\,kg^{-1}$. *(1 mark)*

▲ Figure 2

6. A block of iron of mass 210 g is quickly transferred from a very hot oven into 500 g of water in a beaker. The temperature of water increases from 20 °C to 37 °C.
   specific heat capacity of water = $4200\,J\,kg^{-1}\,K^{-1}$
   specific heat capacity of iron = $450\,J\,kg^{-1}\,K^{-1}$
   a State and explain the transfer of energy between the block and water. *(1 mark)*
   b Calculate the energy gained by the water. *(2 marks)*
   c Estimate the initial temperature of the iron block. Assume there is no transfer of energy to the surroundings. *(3 marks)*

Module 5 Newtonian world and astrophysics
Chapter 14 Thermal physics

In this chapter you will learn about ...

- ☐ Temperature
- ☐ Celsius scale
- ☐ Kelvin scale
- ☐ Thermal equilibrium
- ☐ Kinetic model of matter
- ☐ Brownian motion
- ☐ Internal energy
- ☐ Specific heat capacity
- ☐ Specific latent heat

14 THERMAL PHYSICS
14.1 Temperature
14.2 Solids, liquids, and gases
Specification reference: 5.1.1, 5.1.2

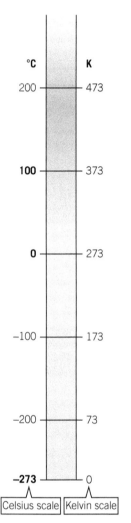

▲ **Figure 1** *Temperatures in °C and K*

Common misconception
Do not confuse heat (thermal) energy and temperature. Energy is measured in joules and temperature in °C or K.

Revision tip: Temperature in K
Thermodynamic temperature in K is always a positive number.

Synoptic link
You will find more information on electrostatic potential energy in Topic 22.5, Electrical potential and energy.

14.1 Temperature

The temperature of a substance is a number which indicates its level of hotness on some chosen scale.

The **Celsius** temperature scale is defined in terms of two accurately reproducible fixed points. The lower fixed point of 0 °C is the temperature of pure melting ice. The upper fixed point of 100 °C is the temperature of steam from boiling water under atmospheric pressure of 1.01×10^5 Pa. The interval between the upper and lower fixed points is divided into 100 equal degrees.

Kelvin scale

The two fixed points for the **thermodynamic temperature scale** are **absolute zero**, 0 K, which is the lowest possible temperature, and the triple point of water, 273.16 K.

0 °C is equivalent to 273.15 K. A change of 1 **kelvin** is equal to a change of 1 degree Celsius, see Figure 1. You can use the equations below to convert temperature between the Celsius and thermodynamic scales.

$$T(\text{K}) \approx \theta(°\text{C}) + 273$$

$$\theta(°\text{C}) \approx T(\text{K}) - 273$$

Thermal equilibrium

When a hot object is in contact with a cooler object, then there is a net flow of thermal energy from the hot object to the cooler object. The temperature of the hot object will decrease and the temperature of the cooler object will increase. Eventually both objects will reach the same temperature.

Two or more objects are in **thermal equilibrium** when there is no net transfer of thermal energy between them.

14.2 Solids, liquids, and gases

The three states or **phases** of matter are solid, liquid, and gas. The **kinetic model** describes how all substances are made up of atoms or molecules.

Kinetic model

In solid and liquid phases, the molecules experience attractive electrostatic force and have negative electrostatic potential energy. The negative simply means that external energy is required to pull apart the molecules. The potential energy is lowest in solids, higher in liquids, and at its highest (0 J) in gases.

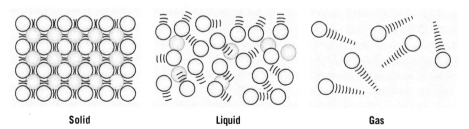

▲ **Figure 2** *Solid, liquid, and gas*

Solid phase

The molecules:

- vibrate about fixed positions and are closely packed together
- have a mixture of kinetic energy (KE) and electrostatic potential energy (PE).

Liquid phase

The molecules:

- can move around each other
- when compared with the solid state:
  - have greater KE
  - have greater mean separation
  - experience smaller attractive electrostatic force
  - have greater PE.

Gas phase

The molecules:

- move freely and rapidly
- collide elastically with each other and move randomly with different speeds in different directions
- have greater mean separation than that in the liquid phase
- experience negligible electrostatic force
- have maximum PE of 0 J.

Brownian motion

Smoke particles in air exhibit random motion because of collisions with air molecules. At a given temperature, the mean kinetic energy of the smoke particles is the same as that of the air molecules. The more massive smoke particles have much smaller speeds than the air molecules.

Synoptic link

The mean kinetic energy of molecules of a gas is directly proportional to the thermodynamic temperature. You will learn more about this in Topic 15.4, The Boltzmann constant.

Summary questions

1. State one difference between temperature and heat (thermal) energy. *(1 mark)*
2. Describe how you could determine 0 °C on an unmarked mercury-in-glass thermometer. *(1 mark)*
3. Convert the following temperatures into K:
   a −270 °C **b** −52 °C **c** 20 °C **d** 10 °C **e** 2000 °C   *(5 marks)*
4. Convert the following temperatures into °C:
   a 3 K **b** 273 K **c** 380 K **d** 500 K **e** 5500 K   *(5 marks)*
5. An ice cube at −10 °C is placed in a warm room. It eventually melts and becomes water at 20 °C.
   Compare the electrostatic potential energy and kinetic energy of the molecules of water at 20 °C with ice at −10 °C.   *(4 marks)*
6. Smoke particles in air are observed to have very small mean speed. Explain how you can deduce that the mean speed of the air molecules is much greater.   *(3 marks)*

14.3 Internal energy
14.4 Specific heat capacity
14.5 Specific latent heat

Specification reference: 5.1.2, 5.1.3

14.3 Internal energy

The internal energy of a substance is defined as the sum of the randomly distributed kinetic and potential energies of atoms or molecules within the substance.

Figure 1 shows the variation of temperature with time for a solid heated at a constant rate. Table 1 summarises what happens to the total energy of the molecules.

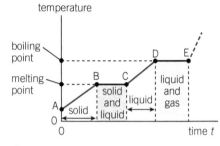

▲ **Figure 1** *Variation of temperature of a substance showing phase changes*

▼ **Table 1**

Section	Kinetic energy	Potential energy	Internal energy
A to B (solid)	Increases	No change	Increases
B to C (phase change from solid to liquid)	No change	Increases	Increases
C to D (liquid)	Increases	No change	Increases
D to E (phase change from liquid to gas)	No change	Increases	Increases

Absolute zero

The **internal energy** of a substance depends on its temperature. At absolute zero, 0 K, all molecules stop moving. The internal energy of the substance is a *minimum* and is entirely due to electrostatic potential energy of the molecules.

14.4 Specific heat capacity

The **specific heat capacity** of a substance is defined as the energy required per unit mass to change the temperature by 1 K (or 1 °C).

The specific heat capacity c of a substance is given by the equation

$$c = \frac{E}{m \Delta \theta} \quad \text{or} \quad E = mc\Delta\theta$$

where E is the energy supplied to the substance, m is the mass of the substance, and $\Delta\theta$ is the change in temperature of the substance. Specific heat capacity has units $J\,kg^{-1}\,K^{-1}$.

Determining specific heat capacity c

The electrical arrangement shown in Figure 2 can be used to determine c of a metal – it can easily be adapted for a liquid. The method of mixtures is illustrated in the worked example.

The energy supplied by the heater in a time t is equal to VIt, where V is the p.d. across the heater and I is the current in the heater. The specific heat capacity of the metal is given by the equation

$$c = \frac{VIt}{m \times [\theta_f - \theta_i]}$$

where m is the mass of the metal block and θ_i and θ_f are the initial and final temperatures of the block.

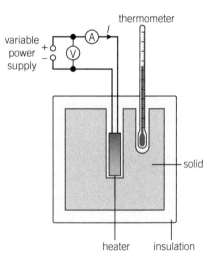

▲ **Figure 2** *Arrangement for determining specific heat capacity of a metal.*

> **Worked example: Specific heat capacity of a liquid**
>
> 100 g of water at 80 °C is poured into 200 g of liquid at 10 °C. The final temperature of the mixture is 50 °C. The specific heat capacity of water is 4200 J kg⁻¹ K⁻¹. Calculate the specific heat capacity c of the liquid. Assume there are no heat losses.
>
> **Step 1:** Calculate the energy transferred to the liquid.
>
> $$E = mc\Delta\theta = 0.100 \times 4200 \times [80 - 50] = 12\,600\,\text{J}$$
>
> **Step 2:** Calculate c.
>
> $$12\,600 = 0.200 \times c \times [50 - 10]$$
>
> $$c = 1575\,\text{J kg}^{-1}\,\text{K}^{-1} \approx 1600\,\text{J kg}^{-1}\,\text{K}^{-1}\,(2\,\text{s.f.})$$

14.5 Specific latent heat

The term latent means 'hidden'.

The **specific latent heat** of a substance L is defined as the energy required to change the phase per unit mass while at constant temperature.

You can use the equation below to calculate the specific latent heat L of a substance.

$$L = \frac{E}{m} \quad \text{or} \quad E = mL$$

where E is the energy supplied to change phase of a mass m of the substance. Specific latent heat has units J kg⁻¹.

Specific latent heat of fusion L_f is used when a substance changes from solid to liquid and **specific latent heat of vaporisation** L_v is used when it changes from liquid to gas.

Determining latent heats

The heater arrangement in Figure 2 can be adapted to determine either L_f or L_v by changing the phase of a substance of mass m in a time t. The energy E supplied by the heater is given by $E = VIt$.

Therefore

$$VIt = mL.$$

> **Summary questions**
>
> **For water:** $c = 4200\,\text{J kg}^{-1}\,\text{K}^{-1}$ $\quad L_f = 3.3 \times 10^4\,\text{J kg}^{-1}$ $\quad L_v = 2.3 \times 10^6\,\text{J kg}^{-1}$
>
> 1 Explain what is meant by absolute zero. *(1 mark)*
> 2 Calculate the energy needed to change 10 g of ice at 0 °C to water at 0 °C. *(2 marks)*
>
> 3 A 200 W heater is used to heat water.
>   How long would it take 300 g of water at 20 °C to reach boiling point? *(3 marks)*
>
> 4 How much longer would it take to completely boil the water in Q3? *(3 marks)*
> 5 A metal block of mass 200 g and temperature 20 °C is placed in 120 g of water initially at 100 °C.
>   The final temperature of the water is 87 °C. Calculate the specific heat capacity c of the metal. *(4 marks)*
> 6 Calculate how long it would take for a 120 W heater to change 500 g of ice at 0 °C into water at 20 °C. *(4 marks)*

Chapter 14 Practice questions

1 Two objects X and Y are in thermal equilibrium. Which statement is correct?

   A There is a net transfer of energy between X and Y.

   B X and Y have the same amount of thermal energy.

   C X and Y have the same temperature in kelvin.

   D X and Y have the same specific heat capacity. *(1 mark)*

2 What is specific latent heat of fusion measured in base units?

   A $m\,s^{-2}$

   B $m^2\,s^{-2}$

   C $m^2\,s^{-2}\,K^{-1}$

   D $kg\,m^2\,s^{-2}$ *(1 mark)*

3 Four solids A, B, C, and D are heated using the same heater. The solids have the same mass.

   A temperature–time graph is plotted for each solid on the same axes (see Figure 1).

   Which solid has the smallest value of specific heat capacity? *(1 mark)*

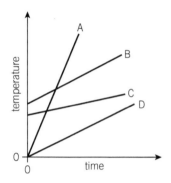

▲ Figure 1

4 A beaker with 120 g of water at 15 °C is placed inside a freezer. It takes 6.0 minutes for the temperature of the water to drop to 0 °C.

   specific heat capacity of water = $4200\,J\,kg^{-1}\,K^{-1}$

   specific latent heat of fusion of ice = $3.3 \times 10^5\,J\,kg^{-1}$

   a Calculate the average rate of energy loss from the water. *(3 marks)*

   b Calculate the time it would take for the water at 0 °C to turn into ice at 0 °C. Assume that the rate of energy transfer is the same as your answer to **a**. *(3 marks)*

5 Figure 2 shows a heater used to melt some crushed ice in a funnel.

   The power of the heater is 25 W. The temperature of the ice remains at 0 °C. The table below shows the variation with time t of the total mass m of the water from the melting ice.

t / s	0	30	60	90	120	150
m / g	0	2.7	5.5	8.2	10.9	14.0

   a Plot a graph of m against t and draw a line of best fit through the data points. *(2 marks)*

   b Use the graph to determine the gradient of the straight line. *(1 mark)*

   c Use your answer to **b** to calculate the specific latent heat of fusion of ice. *(3 marks)*

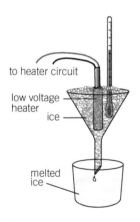

▲ Figure 2

   d Explain why your value the specific latent heat of fusion of ice is smaller than the accepted value of $3.3 \times 10^5\,J\,kg^{-1}$. *(1 mark)*

6 A block of iron of mass 210 g is quickly transferred from a very hot oven into 500 g of water in a beaker. The temperature of water increases from 20 °C to 37 °C.

   specific heat capacity of water = $4200\,J\,kg^{-1}\,K^{-1}$

   specific heat capacity of iron = $450\,J\,kg^{-1}\,K^{-1}$

   a State and explain the transfer of energy between the block and water. *(1 mark)*

   b Calculate the energy gained by the water. *(2 marks)*

   c Estimate the initial temperature of the iron block. Assume there is no transfer of energy to the surroundings. *(3 marks)*

Chapter 15 Ideal gases

In this chapter you will learn about ...

- ☐ Kinetic theory of gases
- ☐ The mole and Avogadro constant
- ☐ Boyle's law
- ☐ Equation of state for ideal gases
- ☐ Absolute zero
- ☐ Root mean square (r.m.s.) speed
- ☐ Boltzmann constant
- ☐ Mean kinetic energy of gas molecules
- ☐ Internal energy of ideal gas

15 IDEAL GASES
15.1 The kinetic theory of gases
15.2 Gas laws
Specification reference: 5.1.4

15.1 The kinetic theory of gases

The kinetic theory of gases is a model used to describe the behaviour of the particles in an **ideal gas**. The assumptions made in the kinetic model for an ideal gas are:

- The gas contains a very large number of atoms (or molecules) moving in random directions with random speeds.
- The volume of the gas atoms is negligible compared with the volume of the gas in the container.
- The atoms collide elastically with each other and with the container walls.
- The time of collision of atoms is negligible compared with the time between the collisions.
- The electrostatic forces between atoms are negligible except during collisions.

The mole and the Avogadro constant

The SI base unit for the **amount of substance** is the **mole**. One mole of *any* substance has 6.02×10^{23} particles (atoms or molecules). The number 6.02×10^{23} is known as the **Avogadro constant**, N_A.

One mole is defined as the amount of substance that contains as many elementary entities as there are atoms in 0.012 kg (12 g) of carbon-12.

You can calculate the total number N of atoms or molecules in a substance using the equation

$$N = n \times N_A$$

where n is the number of moles of the substance.

Explaining pressure

A gas exerts pressure because of repeated elastic collisions of the gas atoms with the container walls.

A gas atom of mass m travelling at a speed u collides with the container wall and bounces back at the same speed u. The change of momentum is $2mu$ (magnitude only). It then travels to the opposite wall and returns back after a time t and makes another collision. The average force f exerted *on* the atom by the wall is given by Newton's second law, $f = \dfrac{2mu}{t}$. According to Newton's third law, the atom exerts an equal but opposite force *on* the wall. The total force F on the wall is from the random collisions of all the atoms inside the container and the pressure p exerted on the wall is $\dfrac{F}{A}$, where A is the area of the wall.

15.2 Gas laws

The relationships between the thermodynamic temperature T, pressure p, and volume V of an ideal gas can be described by a few simple **gas laws**.

Boyle's law

The pressure exerted by a fixed amount of gas is inversely proportional to its volume, provided its temperature remains constant.

This can be expressed mathematically as

$$p \propto \frac{1}{V} \quad \text{or} \quad pV = \text{constant}$$

Pressure, volume, and temperature

Experiments on ideal gases also show that

$$p \propto T \text{ at constant volume}$$

and

$$V \propto T \text{ at constant pressure}$$

Figure 1 shows an arrangement you can use to determine the value of absolute zero in °C. A linear graph is produced by plotting pressure p against temperature θ in °C. The extrapolated graph intersects the temperature axis at absolute zero (−273 °C).

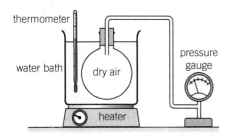
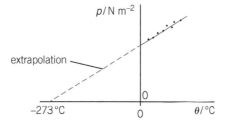

▲ **Figure 1** *Determining absolute zero*

Equation of state for ideal gases

The three relationships above can be combined together to give $\frac{pV}{T}$ = constant.

The 'constant' is equal to nR, where n is the number of moles and R is the **molar gas constant**, and is equal to 8.31 J mol⁻¹ K⁻¹. The **equation of state for an ideal gas** is $pV = nRT$.

> **Revision tip**
> The temperature T is always in kelvin K and not °C.

> **Worked example: Balloon**
>
> Air trapped in a balloon has volume 3.4×10^{-3} m³, temperature 20 °C, and pressure 1.2×10^5 Pa. Calculate the number of air molecules inside the balloon.
>
> **Step 1:** Rearrange the equation of state with n as the subject and calculate n.
>
> $$n = \frac{pV}{RT} = \frac{1.2 \times 10^5 \times 3.4 \times 10^{-3}}{8.31 \times [273 + 20]} = 0.1676$$
>
> **Step 2:** Calculate the number N of air molecules.
>
> $$N = n \times N_A = 0.1676 \times 6.02 \times 10^{23}$$
>
> $$N = 1.0 \times 10^{23} \text{ (2 s.f.)}$$

Summary questions

1. Calculate the number of molecules in 3.0 moles of air. *(1 mark)*
2. The molar mass of oxygen is 0.032 kg mol⁻¹. Calculate the mass of a single molecule. *(2 marks)*

3. The temperature of a gas changes from 100 °C to 200 °C at constant volume. Calculate the factor by which the pressure increases. *(3 marks)*
4. Describe and explain what happens to the volume of a bubble of air rising up through water. *(3 marks)*

5. Calculate the density of 1 mole of air at temperature of −50 °C and pressure of 1.0×10^5 Pa. molar mass of air = 29 g mol⁻¹ *(4 marks)*
6. Estimate the mass of air in a small room. *(4 marks)*

15.3 Root mean square speed
15.4 The Boltzmann constant

Specification reference: 5.1.4

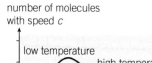

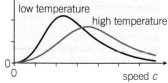

▲ **Figure 1** *Maxwell-Boltzmann distribution showing the spread of speed of the molecules of a gas*

15.3 Root mean square speed

The pressure exerted by a gas and the mean kinetic energy of molecules of a gas are related not to the mean speed of the molecules, but to the root mean square (r.m.s.) speed of the molecules.

r.m.s. speed

The atoms or molecules of a gas move in random directions and have a range of speeds. Figure 1 shows the spread in the speed of molecules – the graph is known as **Maxwell-Boltzmann distribution**.

Consider a container with N atoms. The velocity of each atom is c. The random motion implies that the mean velocity of the atoms is zero.

Here are some important terms:

- mean speed = sum of speed of all the atoms divided by N or simply
$$\bar{c} = \frac{\Sigma \, speed}{N}$$
- mean square speed = sum of velocity² of all the atoms divided by N or
$$\overline{c^2} = \frac{\Sigma c^2}{N}$$
- root mean square speed or r.m.s. speed = $c_{r.m.s.} = \sqrt{\text{mean square speed}}$

For 4 atoms with velocities $-200\,\text{m s}^{-1}$, $+300\,\text{m s}^{-1}$, $-430\,\text{m s}^{-1}$, and $+330\,\text{m s}^{-1}$, you can show that the mean velocity is $0\,\text{m s}^{-1}$, the mean speed is $315\,\text{m s}^{-1}$, and the r.m.s. speed is $325\,\text{m s}^{-1}$.

Pressure at the microscopic level

The kinetic theory model of gases, based on Newtonian mechanics, shows that
$$pV = \tfrac{1}{3} Nm\overline{c^2}$$
where p is the pressure exerted by the gas, V is the volume of the gas, N is the number of atoms, or molecules, in the gas, m is the mass of each atom, and $\overline{c^2}$ is the mean square speed of the atoms. As you see later, this is an important equation because it gives us a better understanding of the mean kinetic energy of the gas atoms and the thermodynamic temperature of the gas.

15.4 The Boltzmann constant

The **Boltzmann constant** k is equal to the molar gas constant R divided by Avogadro constant N_A.

Therefore
$$k = \frac{R}{N_A} = \frac{8.31}{6.02 \times 10^{23}} = 1.38 \times 10^{-23}\,\text{J K}^{-1}$$

The equation of state $pV = nRT$, can also be written as
$$pV = n(kN_A)T$$
or
$$pV = NkT$$

where N is the number of atoms or molecules in the gas. Be careful – the lower case n stands for the 'number of moles of gas' and the capital N represents the 'number of atoms or molecules'.

> **Maths: Mathematical symbols**
>
> The sigma Σ symbol stands for 'sum of ...' and the 'bar' on top of quantities stands for 'mean of ...'.

Ideal gases

Mean kinetic energy and temperature

The proof below shows the derivation of an important equation ($\frac{1}{2}m\overline{c^2} = \frac{3}{2}kT$) – it relates the thermodynamic temperature of the gas to the mean kinetic energy of the atoms or molecules.

Main equations: $pV = nRT$, $pV = \frac{1}{3}Nm\overline{c^2}$, $N = nN_A$, and $R = kN_A$

$$pV = \tfrac{1}{3}Nm\overline{c^2} = nRT$$

Therefore

$$\tfrac{1}{3}m\overline{c^2} = \frac{nRT}{N} = \frac{RT}{N_A} = kT$$

This can be written as

$$\tfrac{1}{2}m\overline{c^2} = \tfrac{3}{2}kT$$

In the equation above, $\frac{1}{2}m\overline{c^2}$ is the mean kinetic energy of the gas molecules.

Note:

At a particular temperature T:
- all gas atoms have the same mean kinetic energy
- the greater mass atoms have smaller r.m.s. speed.

Internal energy

The internal energy of an ideal gas is entirely in the form of kinetic energy. The potential energy is zero because of the negligible electrostatic forces between atoms. Therefore

internal energy of a gas = $N \times \frac{3}{2}kT$

or

internal energy $\propto T$

> **Worked example: Air molecules**
>
> Calculate the r.m.s. speed of oxygen molecules in a room at a temperature of 20 °C.
> mass of oxygen molecule = 5.3×10^{-26} kg
>
> **Step 1:** Derive the equation for the r.m.s. speed.
>
> $\tfrac{1}{2}m\overline{c^2} = \tfrac{3}{2}kT$
>
> mean square speed = $\overline{c^2} = \dfrac{3kT}{m}$
>
> r.m.s. speed = $c_{r.m.s.} = \sqrt{\dfrac{3kT}{m}}$
>
> **Step 2:** Substitute values to determine the r.m.s. speed.
>
> $c_{r.m.s.} = \sqrt{\dfrac{3 \times 1.38 \times 10^{-23} \times (273 + 20)}{5.3 \times 10^{-26}}} = 480 \text{ m s}^{-1}$ (2 s.f.)

Revision tip
Mean kinetic energy of molecules is proportional to the thermodynamic temperature T.

Summary questions

1. The velocities of 5 atoms in m s^{-1} are −100, −200, 150, 200, and 300. Calculate the mean velocity and the mean speed of these atoms. *(2 marks)*

2. Calculate the mean square speed and the r.m.s. speed of the atoms in **Q1**. *(2 marks)*

3. Calculate the mean kinetic energy of gas molecules at a temperature of 200 °C. *(2 marks)*

4. In the demonstration of Brownian motion with smoke particles in air, explain why the smoke particles have a much smaller speed than the air molecules. *(2 marks)*

5. Calculate the r.m.s. speed of the molecules in **Q3** given the mass of each molecule is 4.8×10^{-26} kg. *(3 marks)*

6. Calculate the internal energy of 2.0 moles of gas atoms at a temperature of 0 °C. *(3 marks)*

Chapter 15 Practice questions

1 Which statement is correct about absolute zero?
  The atoms of a substance:
  A have no total energy
  B have no kinetic energy
  C have no internal energy
  D have no potential energy (1 mark)

2 The molar mass of carbon is $12\,\text{g mol}^{-1}$.
  How many atoms of carbon are there in 1.0 kg?
  A 5.0×10^{22}
  B 7.2×10^{22}
  C 7.2×10^{24}
  D 5.0×10^{25} (1 mark)

3 The actual velocity of four atoms in m s^{-1} is +200, +300, −500, and +600.
  What is the r.m.s. speed of the atoms?
  A $180\,\text{m s}^{-1}$
  B $280\,\text{m s}^{-1}$
  C $430\,\text{m s}^{-1}$
  D $450\,\text{m s}^{-1}$ (1 mark)

4 The r.m.s. speed of gas atoms is $500\,\text{m s}^{-1}$ at $50\,°\text{C}$. The temperature of the gas is doubled to $100\,°\text{C}$.
  What is the r.m.s. speed of the atoms at this higher temperature?
  A $540\,\text{m s}^{-1}$
  B $580\,\text{m s}^{-1}$
  C $710\,\text{m s}^{-1}$
  D $1000\,\text{m s}^{-1}$ (1 mark)

5 A scientist is conducting an experiment to determine the Boltzmann constant k using 1.0 mole of gas trapped in a container of volume $2.2 \times 10^{-2}\,\text{m}^2$. Figure 1 shows a graph of pressure p exerted by the gas against temperature T in kelvin.
  a Write an equation relating pressure p exerted by the gas, its volume V, absolute temperature T, and the molar gas constant R. (1 mark)
  b Explain why the graph shown in Figure 1 is a straight-line graph. (1 mark)
  c Use Figure 1 to determine a value for k. (3 marks)

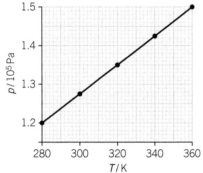

▲ Figure 1

6 A container of volume $5.2 \times 10^{-4}\,\text{m}^3$ has 0.65 g of trapped air. The air pressure inside the container is $1.2 \times 10^5\,\text{Pa}$. The molar mass of air is about $30\,\text{g mol}^{-1}$.
  Calculate:
  a the density of the air in the container; (1 mark)
  b the number of moles of air in the container; (1 mark)
  c the temperature in °C of the air in the container; (3 marks)
  d the mean kinetic energy of the molecules in the container. (2 marks)

Chapter 16 Circular motion

In this chapter you will learn about ...

- [ ] Angular velocity
- [ ] Radians
- [ ] Circular motion
- [ ] Centripetal force
- [ ] Centripetal acceleration

16 CIRCULAR MOTION
16.1 Angular velocity and the radian
16.2 Centripetal acceleration

Specification reference: 5.2.1, 5.2.2

16.1 Angular velocity and the radian

There are many examples of objects moving in circular paths – a moon orbiting a planet, a car on a roundabout, an electron in a uniform magnetic field, and so on. It is easier to analyse the motion of such objects in terms of their angular velocity.

The radian

Angles can be measured in degrees. There are 360° in a full circle. Angles can also be measured in radians.

A **radian** is the angle subtended by a circular arc with a length equal to the radius of the circle.

An angle of 1 radian, or 1 rad, is about 57°. In general, the angle θ in radians is defined as follows:

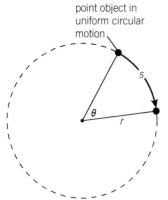

▲ **Figure 1** *The angle θ in radians is given by the equation $\theta = \dfrac{s}{r}$*

$$\text{angle (in radians)} = \frac{\text{arc length}}{\text{radius of circle}} = \frac{s}{r} \text{ (see Figure 1)}$$

For a complete circle, s = circumference = $2\pi r$. Therefore, 360° is the same as 2π rad. You can also show that $180° = \pi$, $90° = \dfrac{\pi}{2}$, and so on.

Angular velocity ω

The **angular velocity** of an object moving in a circle is defined as the rate of change of angle.

The equation for angular velocity ω is

$$\omega = \frac{\theta}{t}$$

where θ is the angle through which an object moves in a time t. Angular velocity is measured in radians per second, rad s^{-1}. In a time t equal to one period T, the object will move through an angle θ equal to 2π radians. Therefore

$$\omega = \frac{2\pi}{T}$$

> **Revision tip**
> To avoid confusing the terms **velocity** and **angular velocity**, just remember that velocity has units **m s^{-1}** and angular velocity is measured in **rad s^{-1}**.

The frequency f of the rotating object is given by $f = \dfrac{1}{T}$. Therefore, we can also use the following equation for angular velocity:

$$\omega = 2\pi f$$

16.2 Centripetal acceleration

Figure 2 shows an object moving in a circle at a constant speed v. The velocity of the object is continually changing because its direction of travel changes. The object must be accelerating because acceleration is defined as the rate of change of velocity.

Speed in a circle

You can determine the speed v of an object moving in a circle of radius r at a constant angular velocity ω as follows:

$$v = \frac{\text{circumference}}{\text{period}}$$

$$v = \frac{2\pi r}{T}$$

However, $\omega = \dfrac{2\pi}{T}$, therefore

$$v = \omega r$$

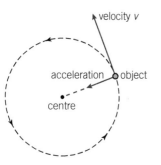

▲ **Figure 2** *An object moving in a circle has acceleration*

Circular motion

Centripetal force and acceleration

For an object travelling in a circle of radius r at a constant speed v, it has a constant acceleration a towards the centre of the circle. This acceleration is known as **centripetal acceleration**. Centripetal simple means 'towards the centre'. The equation for the centripetal acceleration is

$$a = \frac{v^2}{r}$$

Since $v = \omega r$, we can also use the equation

$$a = \omega^2 r$$

According to $F = ma$, the resultant force on the object and the acceleration must be in the same direction – towards the centre of the circle. This resultant force is referred to as **centripetal force**.

Model answer: Mercury

The planet Mercury orbits the Sun in 88 days in a circle of radius 5.8×10^{10} m. Calculate its centripetal acceleration.

Answer

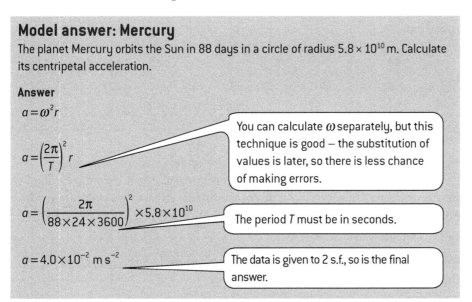

$a = \omega^2 r$

$a = \left(\frac{2\pi}{T}\right)^2 r$

You can calculate ω separately, but this technique is good – the substitution of values is later, so there is less chance of making errors.

$a = \left(\frac{2\pi}{88 \times 24 \times 3600}\right)^2 \times 5.8 \times 10^{10}$

The period T must be in seconds.

$a = 4.0 \times 10^{-2}$ m s^{-2}

The data is given to 2 s.f., so is the final answer.

Summary questions

1 Change the following angles into radians:
   a 45°   b 5.0°   c 420°   *(3 marks)*
2 Calculate the speed of an object travelling in a circle of radius 50 cm with an angular velocity of 15 rad s^{-1}. *(2 marks)*
3 An object travels through an angle of 45° in 2.5 s. Calculate its angular velocity in rad s^{-1}. *(2 marks)*
4 An aircraft moves in a circular path of radius 12 km at a constant speed of 150 m s^{-1}.
   Calculate the angular velocity and centripetal acceleration of the aircraft. *(4 marks)*
5 A car travelling at 20 m s^{-1} over a humpback bridge has acceleration equal to the acceleration of free fall. Calculate the radius of curvature of the bridge. *(2 marks)*
6 Explain why the speed of an object moving in a circle is not affected by the centripetal force. *(2 marks)*

16.3 Exploring centripetal forces

Specification reference: 5.2.2

> **Revision tip**
> In the equations $F = \dfrac{mv^2}{r}$ and $F = m\omega^2 r$, F is the *total* or *resultant* force acting on the object which must *point towards the centre of the circle*.

16.3 Exploring centripetal forces

For an object (planet, car, aeroplane, etc.) moving in a circle of radius r at a constant speed v, the centripetal acceleration a can be calculated using either

$$a = \frac{v^2}{r}$$

or

$$a = \omega^2 r$$

where ω is the angular velocity. The resultant force on the object, which is the **centripetal force** F, can be calculated using Newton's second law $F = ma$. Therefore

$$F = \frac{mv^2}{r}$$

and

$$F = m\omega^2 r$$

Investigating centripetal forces

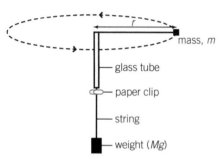

▲ **Figure 1** *Whirling bung experiment*

Figure 1 shows an arrangement you can use to demonstrate and analyse circular motion. A rubber bung of mass m is whirled in a horizontal circle at a constant speed. The speed v can be measured directly using a motion sensor. Alternatively, the time t for N revolutions could be timed using a stopwatch and v calculated using $v = \dfrac{2prN}{t}$.

For this arrangement, the tension in the string provides the centripetal force. Therefore

$$F = \frac{mv^2}{r} = Mg$$

The radius r of the circle could be kept constant and different values obtained for v as m is changed.

A graph of v^2 against m should be a straight line passing through the origin.

Car on a roundabout

Figure 2 shows a car moving round a roundabout.

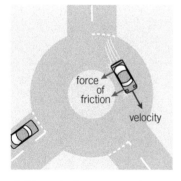

▲ **Figure 2** *Car on a roundabout*

The centripetal force is provided by the frictional force F_R between the tyres and the road. Therefore

$$F_R = \frac{mv^2}{r}$$

Vertical circle

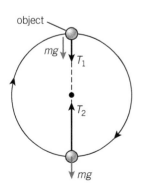

▲ **Figure 3** *Motion in a vertical circle*

Figure 3 shows an object tied to a string whirled in a vertical circle of constant radius r and a constant speed v. The centripetal force on the object must be constant and equal to $\dfrac{mv^2}{r}$. This can only happen if the tension in the string changes.

At the top of the motion: centripetal force = weight + tension = $mg + T_1$

Therefore

$$mg + T_1 = \frac{mv^2}{r}$$

$$T_1 = \frac{mv^2}{r} - mg$$

At the bottom of the motion: centripetal force = tension − weight = $T_2 - mg$

Therefore

$$T_2 - mg = \frac{mv^2}{r}$$

$$T_2 = \frac{mv^2}{r} + mg$$

The tension in the string is greatest at the bottom.

Banked turn

Figure 4 shows an aeroplane turning or banking while flying horizontally. The lift force L is at an angle to the vertical. This force has both vertical and horizontal components. The weight W of the aeroplane acts vertically and has no horizontal component.

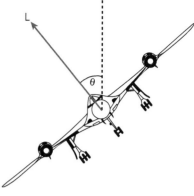

▲ **Figure 4** *An aeroplane turning*

Vertically: Resultant force = 0

vertical component of L = weight

$$L \cos\theta = W$$

Horizontally: The horizontal component of L provides the centripetal force.

$$L \sin\theta = \frac{mv^2}{r}$$

> **Worked example: Whirling bung**
>
> An object of mass 50 g is attached to a light rod of length 10 cm and spun at a constant speed of 3.0 m s^{-1} in a *vertical* circle. Calculate the maximum tension in the rod.
>
> **Step 1:** Calculate the centripetal force on the object.
>
> $$F = \frac{mv^2}{r} = \frac{0.050 \times 3.0^2}{0.10} = 4.5 \text{ N}$$
>
> **Step 2:** The maximum tension T is at the bottom of the motion. Write an equation for the tension and then substitute the values.
>
> $T -$ weight $= 4.5$
>
> $T - (0.050 \times 9.81) = 4.5$
>
> $T = 5.0$ N (2 s.f.)

Summary questions

1. Suggest what provides the centripetal force on a planet orbiting the Sun. *(1 mark)*

2. Calculate the centripetal force on a stone of mass 0.20 kg that is whirled in a horizontal circle of radius 0.50 m at a speed of 3.2 m s^{-1}. *(2 marks)*

3. A car of mass 900 kg is travelling at a constant speed round a roundabout of radius 30 m. The frictional force provided by the tyres is 4.3 kN. Calculate the speed of the car. *(2 marks)*

4. In the whirling bung experiment, what is the gradient of the v^2 against m graph equal to? *(2 marks)*

5. The Moon has mass 7.3×10^{22} kg and is at a distance of 3.8×10^8 m from the Earth. It takes 27 days to orbit the Earth. Calculate the gravitational force acting on the Moon. *(3 marks)*

6. Show that $\tan\theta = \dfrac{v^2}{rg}$ for an aeroplane turning (see Figure 4). *(3 marks)*

Chapter 16 Practice questions

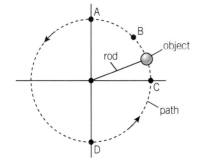
▲ Figure 1

1 What is the correct unit for angular velocity?
   A m s^{-1}
   B Hz
   C rad
   D rad s^{-1} *(1 mark)*

2 An object is attached to a rod of negligible mass, see Figure 1. The object is rotated at a constant speed in a vertical circle. At which position is the tension in the rod maximum? *(1 mark)*

3 An object moves in a circular path of radius 10 cm with a constant speed of 2.6 m s^{-1}.
   What is the frequency of the object moving in the circle?
   A 0.24 Hz
   B 4.1 Hz
   C 26 Hz
   D 160 Hz *(1 mark)*

4 An object is moving in a circle. The centripetal force on the object is F. The speed of the object is now doubled and its radius halved.
   What is the new value for the centripetal force in terms of F?
   A F
   B $2F$
   C $4F$
   D $8F$ *(1 mark)*

5 a An object is moving in a circle at a constant speed. Explain why the object must have acceleration. *(2 marks)*
   b Figure 2 shows an object of mass 120 g on a revolving metal disc. The object moves in a circle of radius 10 cm.
      i On Figure 2, show the direction of the frictional force acting on the object as it travels round in a circle. *(1 mark)*
      ii The maximum frictional force acting on the object is equal to half the weight of the object. The speed of the disc is slowly increased. Calculate the maximum speed of the object for it to remain on the disc. *(4 marks)*

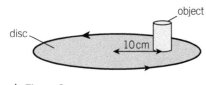
▲ Figure 2

6 An electron moves in a circle of radius 1.2 cm at a constant speed of 4.2×10^7 m s^{-1}.
   Calculate:
   a the centripetal acceleration of the electron; *(2 marks)*
   b the centripetal force on the electron; *(2 marks)*
   c the period of the electron as it moves round in a circle. *(2 marks)*

7 Figure 3 shows a conical pendulum where an object of weight 1.2 N attached to the end of a string describes a horizontal circle.
   The radius of the circular path is 15 cm and the string makes an angle of 45° to the vertical.
   a Show that the tension in the string is 1.7 N. *(1 mark)*
   b Show that the centripetal force acting on the object is 1.2 N. *(1 mark)*
   c Calculate the speed v of the object. *(3 marks)*

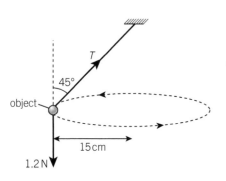
▲ Figure 3

Chapter 17 Oscillations

In this chapter you will learn about ...

- ☐ Simple harmonic motion
- ☐ Phase difference
- ☐ Period and frequency of oscillations
- ☐ Graphs for simple harmonic motion
- ☐ Energy of a simple harmonic oscillator
- ☐ Damping
- ☐ Resonance

17 OSCILLATIONS
17.1 Oscillations and simple harmonic motion

Specification reference: 5.3.1

17.1 Oscillations and simple harmonic motion

A swinging pendulum, an oscillating mass at the end of a vertical spring, and a vibrating ruler fixed at one end are all examples of free oscillations. Simple harmonic motion (SHM) is the simplest type of motion of oscillators. One important characteristic of SHM is that the period of the oscillations is independent of the amplitude of the oscillator. The oscillator keeps 'steady time' and is known as an **isochronous oscillator**. (In Greek, *iso = same* and *chronous = time*.)

Oscillatory motion

Figure 1 shows the displacement against time graph for a simple harmonic oscillator.

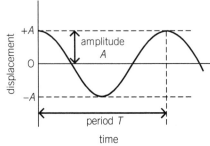

▲ **Figure 1** *The displacement–time graph for an oscillator*

Here are some important definitions:

- The **displacement** x is the distance from the equilibrium position in a particular direction.
- The **amplitude** A is the maximum displacement of the oscillator from its equilibrium position.
- The **period** T is the time taken for the oscillator to complete one complete oscillation.
- The **frequency** f is the number of oscillations per unit time. Frequency is measured in hertz, Hz.

Simple harmonic motion

An oscillator executes **simple harmonic motion** when its acceleration is directly proportional to its displacement from its **equilibrium position**, and is directed towards the equilibrium position.

For such an oscillator

$$\text{acceleration} \propto -\text{displacement}$$

or

$$a \propto -x$$

The minus sign implies that the direction of the acceleration is always towards a fixed point (equilibrium position), see Figure 2. The constant of proportionality is ω^2, where ω is the angular frequency of the oscillator.

The angular frequency ω is defined by either the equation $\omega = 2\pi f$ or $\omega = \dfrac{2\pi}{T}$.

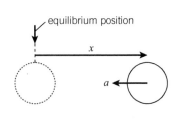

▲ **Figure 2** *Acceleration and displacement are in opposite directions*

> **Common misconception: Omega?**
>
> ω is the **angular** *velocity* of the object moving in a *circle*. In oscillatory motion, ω is the **angular** *frequency* of the oscillator. The equations and unit for ω are the same for both. Use the terms carefully.

Oscillations

Phase difference

Phase difference is a useful quantity when comparing the motions of two simple harmonic oscillators with identical period of oscillations.

The **phase difference** is the fraction of an oscillation one oscillator leads or lags behind another.

Phase difference, as for waves, can be measured either in degrees or radians. Figure 3 illustrates the motions of two oscillators A and B. The oscillator A *lags* behind oscillator B by a time Δt.

The phase difference ϕ in radians between the motions is given by the equation

$$\phi = 2\pi \times \left(\frac{\Delta t}{T}\right) \text{ radians}$$

where T is the common period of the two oscillators.

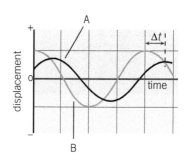

▲ **Figure 3** *There is a phase difference between the motions of oscillators A and B*

Investigationing period and frequency

The period T of a simple harmonic oscillator can be determined by measuring the time t taken for N oscillations; $T = \frac{t}{N}$. Having a large number of oscillations reduces the percentage uncertainty due to reaction time.

You can also use a motion sensor connected to a laptop or a data-logger to show that an oscillator keeps steady time (isochronous).

> ### Worked example: Oscillating pendulum
>
> A pendulum bob of mass 120 g oscillates with a period of 1.40 s and has amplitude of 50.0 cm.
>
> Calculate the maximum resultant force acting on the pendulum bob.
>
> **Step 1:** Derive an equation for the force F in terms of the quantities given.
>
> $F = ma$, and $a_{max} = \omega^2 A$ with $\omega = \frac{2\pi}{T}$
>
> Therefore, the maximum force $F_{max} = m\omega^2 A = \frac{4\pi^2 mA}{T^2}$
>
> **Step 2:** Calculate the value for the maximum force.
>
> $m = 0.120 \text{ kg} \qquad A = 0.500 \text{ m} \qquad T = 1.40 \text{ s}$
>
> $F_{max} = \frac{4\pi^2 \times 0.120 \times 0.500}{1.40^2} = 1.21 \text{ N (3 s.f.)}$

Summary questions

1. Calculate the angular frequency of an oscillator with period 2.00 s. *(1 mark)*
2. Describe how the acceleration of an oscillator changes as it travels from its equilibrium position to its maximum displacement. *(2 marks)*
3. An oscillator with frequency 1.5 kHz has amplitude 0.60 mm. Calculate the angular frequency and maximum acceleration of the oscillator. *(4 marks)*
4. Calculate the phase difference between the oscillators A and B shown in Figure 3 when the lag time is one-eighth of the period. *(1 mark)*
5. The acceleration of an oscillator is given by the equation $a = -400x$. Calculate its period. *(3 marks)*
6. The period T of a pendulum of length L is given by $T = 2\pi \left(\frac{L}{g}\right)^{\frac{1}{2}}$, where g is the acceleration of free fall.

   A pendulum has length 5.0 m. The mass of the bob is 1.2 kg and the amplitude is 1.6 m. Calculate the maximum resultant force on the oscillating pendulum bob. *(4 marks)*

17.2 Analysing SHM
17.3 SHM and energy

Specification reference: 5.3.1, 5.3.2

17.2 Analysing SHM

The variation of the displacement x of a simple harmonic with time t is sinusoidal. This simply means that x against t is a sine or cosine shaped graph. You have already seen this in Figure 1 in Topic 17.1, Oscillations and simple harmonic motion.

Displacement equations

You need to be familiar with the following two equations for displacement x:

$$x = A \sin \omega t \text{ and } x = A \cos \omega t$$

The amplitude of the oscillations is A and ω is the angular frequency. These equations are 'solutions' of the SHM equation $a = -\omega^2 x$.

- Use the sine version when $x = 0$ at $t = 0$.
- Use the cosine version when $x = A$ at $t = 0$.

Revision tip
Make sure your calculator is in *radian mode* when using the equations $x = A \cos \omega t$ and $x = A \sin \omega t$.

Graphs

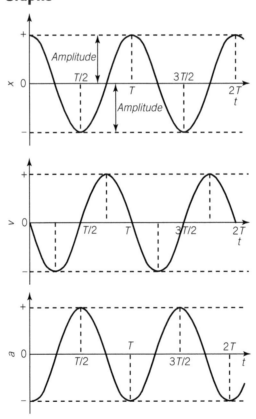

▲ **Figure 1** *Simple harmonic graphs*

Synoptic link
You learnt about displacement–time graphs and velocity–time graphs in Chapter 3, Motion.

The velocity v of the oscillator can be determined from the *gradient* of the displacement–time graph and the acceleration a from the *gradient* of the velocity–time graph. Figure 1 show x–t, v–t, and a–t graphs for a simple harmonic oscillator.

Velocity equations

Imagine a swinging pendulum. When displacement $x = 0$ (equilibrium position), the pendulum bob has maximum speed and the speed decreases as its displacement increases. The speed is momentarily zero when $x = A$. You can

determine the velocity v of any simple harmonic oscillator using the equation
$v = \pm \omega \sqrt{A^2 - x^2}$

The maximum speed is when $x = 0$ and is given by the equation $v_{max} = \omega A$.

17.3 SHM and energy

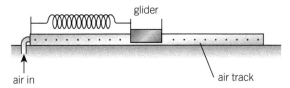

▲ **Figure 2** *A spring–mass oscillator has kinetic energy and potential energy*

The energy of an oscillator is made up of kinetic energy and potential energy. For example, a vibrating atom in a solid has kinetic energy because of its motion and electrostatic potential energy due to the electrical forces on the atom. For a spring–mass system (see Figure 2) oscillating horizontally, the stored energy is elastic potential energy in the spring.

Energy–displacement graph

The kinetic energy E_k of a simple harmonic oscillator of mass m is given by

$$E_k = \tfrac{1}{2} mv^2 = \tfrac{1}{2} m\omega^2 (A^2 - x^2)$$

- E_k has a maximum value of $\tfrac{1}{2} m\omega^2 A^2$ when $x = 0$.
- The total energy of the oscillator must also be $\tfrac{1}{2} m\omega^2 A^2$.
- The potential energy E_p of the oscillator must be $\tfrac{1}{2} m\omega^2 x^2$.

Figure 3 shows the variation of E_k and E_p with displacement x.

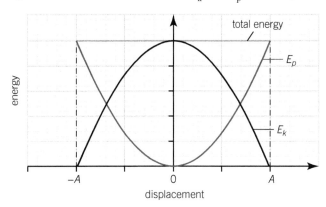

▲ **Figure 3** *The variation of potential and kinetic energies with displacement*

🖩 Worked example: Energy of an oscillating mass

A 700 g mass is hung from a spring. The mass is pulled vertically below its equilibrium and then released. It oscillates with a period of 0.75 s and amplitude of 32 cm.

Calculate the maximum kinetic energy E_{max} of the mass.

Step 1: Calculate the angular frequency ω.

$$\omega = \frac{2\pi}{T} = \frac{2\pi}{0.75} = 8.379 \text{ rad s}^{-1}$$

Step 2: The maximum speed v_{max} is when the mass is at its equilibrium. Calculate v_{max}.

$$v_{max} = \omega A = 8.379 \times 0.32 = 2.68 \text{ m s}^{-1}$$

Step 3: Calculate E_{max}.

$$E_{max} = \tfrac{1}{2} mv^2 = \tfrac{1}{2} \times 0.700 \times 2.68^2$$

$$E_{max} = 2.5 \text{ J (2 s.f.)}$$

Summary questions

1. State how the maximum speed of an oscillator is related to the amplitude of its motion. *(1 mark)*

2. Explain how the first and last graphs in Figure 1 illustrate simple harmonic motion. *(2 marks)*

3. The oscillations of a vibrating plate have amplitude 3.0 mm and period 40 ms. Calculate the maximum speed of the vibrating plate. *(2 marks)*

4. For the plate in Q3, calculate its speed when the displacement of the plate is 1.0 mm. *(2 marks)*

5. An oscillator of mass 180 g has period of 0.80 s and amplitude 20 cm. Calculate the potential energy of the oscillator when the displacement is 10 cm. *(3 marks)*

6. The oscillator in Q5 has maximum displacement at time $t = 0$. Calculate the time t when its displacement is 7.5 cm for the first instant. *(2 marks)*

17.4 Damping and driving
17.5 Resonance
Specification reference: 5.3.3

17.4 Damping and driving

The motion of a mechanical system oscillating without any external forces is known as **free oscillations**. The frequency of the free oscillations is known as the **natural frequency** of the oscillator. In **forced oscillations**, an external periodic driving force is applied to an oscillator. The frequency of the driving force is known as the **driving frequency**.

Oscillations of a mechanical system are damped when an external force acts on the system which reduces the amplitude of the oscillations.

Damping

Damping is caused by frictional forces, for example air resistance and viscous drag in oil. For a mechanical system such as a car, damping is essential because without it, the car would keep bouncing up and down every time it went over a bump.

Figure 1 shows how the displacement of a damped oscillator varies with time. For **light damping**, the amplitude of the oscillator decreases by the same fraction every oscillation – the amplitude is said to decay exponentially.

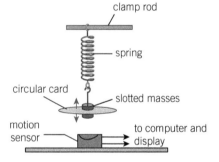

▲ **Figure 2** *Investigating damped oscillations using a motion sensor*

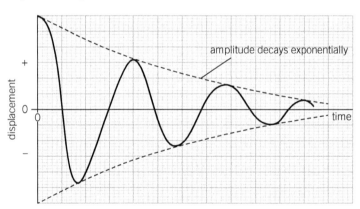

▲ **Figure 1** *Damped oscillations of an oscillator – the period remains the same*

The damped oscillations of a spring–mass system can be investigated using the arrangement shown in Figure 2.

17.5 Resonance

Figure 3 shows an arrangement that may be used to investigate the forced oscillations of a mechanical oscillator. As the driver frequency is slowly increased from zero, the amplitude of oscillations of the forced spring–mass system increases until it reaches maximum amplitude at a particular frequency, and then the amplitude decreases again.

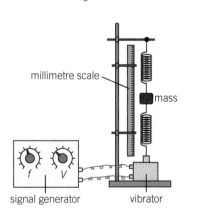

▲ **Figure 3** *Investigating forced oscillations and resonance*

Oscillations

Characteristic of resonance

The oscillator is in resonance when it has maximum amplitude when being forced to oscillate.

At resonance:
- the driving frequency is equal to the natural frequency f_0 of the forced oscillator (for light damped system)
- the forced oscillator absorbs maximum energy from the driver.

The frequency at the maximum amplitude is also known as the resonant frequency.

Figure 4 shows the variation of amplitude of the forced oscillator with the driving frequency for different degrees of damping. As the amount of damping is increased:

- the amplitude of vibration at any frequency decreases
- the peak on the graph becomes flatter and broader
- the maximum amplitude occurs at a lower frequency than f_0.

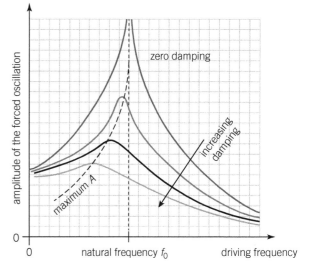

▲ Figure 4 Resonance curves

Some examples of resonance: tuning circuit of a radio, MRI scanner, and wind instrument (e.g., organ pipe).

> ### 🖩 Worked example: Resonance
>
> Figure 5 shows a resonance curve drawn by a student. The values indicated on the axes are correct, but the labelling has been omitted.
>
> The mass of the oscillator is 100 g.
>
> State the labelling for the axes and determine the total energy of the oscillator at resonance.
>
> **Step 1:** Correctly identify the labels for the axes.
>
> The x-axis should be 'driving frequency/Hz' and the y-axis should be 'amplitude/cm'.
>
> **Step 2:** Determine the angular frequency ω.
>
> $\omega = 2\pi f = 2\pi \times 4.0 = 25.1 \text{ rad s}^{-1}$
>
> **Step 3:** Calculate the total energy.
>
> The total energy of the oscillator is the maximum kinetic energy.
>
> total energy $= \frac{1}{2} m \omega^2 A^2 = \frac{1}{2} \times 0.100 \times 25.1^2 \times 0.020^2$
>
> total energy $= 0.013$ (2 s.f.)

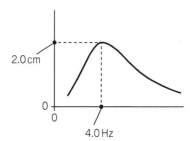

▲ Figure 5

Summary questions

1. State what is meant by resonance. *(1 mark)*
2. State what is meant by damped oscillations. *(1 mark)*

3. Describe how the 'sharpness' of the resonance curve is affected by the degree of damping. *(1 mark)*
4. Describe how the resonant frequency of a specific oscillator is affected by the degree of damping. *(1 mark)*

5. The amplitude of an oscillator decays exponentially. The amplitude decreases by half after every 10 oscillations. How many oscillations does it take for the amplitude to become an eighth of its initial amplitude? *(2 marks)*
6. The oscillator illustrated in the worked example is set up with less damping. Resonance occurs at the same frequency as before but the amplitude increases to 20 cm. Calculate the total energy of the oscillator under these conditions. *(3 marks)*

Chapter 17 Practice questions

1. What is the correct unit for angular frequency?

   A m s^{-1} C rad

   B Hz   D rad s^{-1} *(1 mark)*

2. The maximum speed of a simple harmonic oscillator is v_{max} when it oscillates with amplitude A and period T.

   What is the correct equation for v_{max}?

   A $v_{max} = \dfrac{A}{T}$

   B $v_{max} = \dfrac{2A}{T}$

   C $v_{max} = \dfrac{\pi A}{T}$

   D $v_{max} = \dfrac{2\pi A}{T}$ *(1 mark)*

3. The period of oscillations of a simple harmonic oscillator is T. The amplitude of the oscillator is now doubled.

   What is the new value of its period in terms of T?

   A $\dfrac{T}{2}$

   B T

   C $\sqrt{2}T$

   D $2T$ *(1 mark)*

4. The acceleration a of a simple harmonic oscillator is related to its displacement x by the equation $a = -100x$.

   What is the frequency of the oscillations?

   A 0.63 Hz   C 63 Hz

   B 1.6 Hz   D 100 Hz *(1 mark)*

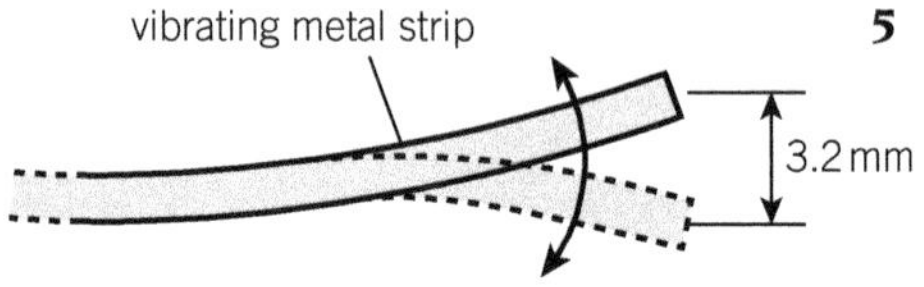

5. Figure 1 shows the end of a vibrating metal strip.

   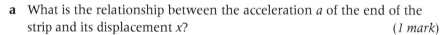

   ▲ Figure 1

   The metal strip oscillates with a simple harmonic motion with a frequency of 840 Hz. The distance between the two extreme positions of the end of the strip is 3.2 mm.

   a What is the relationship between the acceleration a of the end of the strip and its displacement x? *(1 mark)*

   b State the amplitude of the oscillating end of the strip. *(1 mark)*

   c Calculate the maximum:

       i velocity of the end of the strip; *(2 marks)*

       ii acceleration of the end of the strip. *(2 marks)*

   d Describe how the potential energy of the strip varies as the end of the strip travels from the equilibrium position to its amplitude. *(2 marks)*

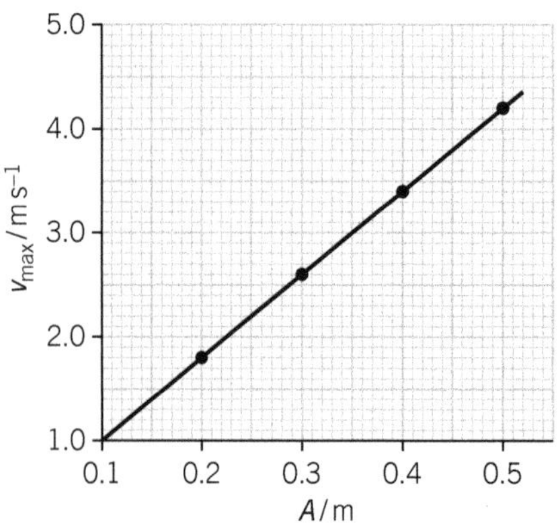

6. A student is investigating the oscillations of a trolley connected to springs. The maximum speed v_{max} of the trolley as it passes through the equilibrium position is measured and recorded using a data-logger.

   Figure 2 shows the variation of v_{max} with the amplitude A of the oscillations.

   a Write an equation for v_{max} in terms of A. *(1 mark)*

   b Explain why the graph shown in Figure 2 is a straight line. *(1 mark)*

   c i Determine the gradient of the line shown in Figure 2. *(1 mark)*

       ii Calculate the frequency f of the oscillations. *(3 marks)*

▲ Figure 2

Chapter 18 Gravitational fields

In this chapter you will learn about ...

- ☐ Gravitational field lines
- ☐ Gravitational field strength
- ☐ Newton's law of gravitation
- ☐ Radial gravitational field
- ☐ Kepler's laws
- ☐ Motion of planets and satellites
- ☐ Geostationary orbits
- ☐ Gravitational potential
- ☐ Gravitational potential energy
- ☐ Escape velocity

18 GRAVITATIONAL FIELDS
18.1 Gravitational fields
18.2 Newton's law of gravitation
18.3 Gravitational field strength for a point mass

Specification reference: 5.4.1, 5.4.2

18.1 Gravitational fields

All objects have mass and they all create a **gravitational field** in the space around them. A mass placed in the gravitational field of another object will experience *attractive* gravitational force. For example, the Earth creates a gravitational field and a person in this field will experience a gravitational force – we call this the weight of the person.

Field patterns

Gravitational field patterns can be mapped using **gravitational field lines** (or lines of force). The direction of the gravitational field is indicated by an arrow on the field lines. The **gravitational field strength** (see later) is indicated by the separation between the field lines (see Figure 1).

- A **uniform field** has equally spaced field lines.
- A **radial field** has straight field lines converging to a point at the centre of the object. The field strength gets smaller with increased distance from the object.

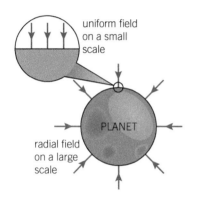

▲ **Figure 1** *Radial and uniform gravitational fields*

Gravitational field strength g

The gravitational field strength is the gravitational force experienced per unit mass on a small object at that point.

This can be written as

$$g = \frac{F}{m}$$

where F is the gravitational force on the object of mass m. The unit for g is $N\,kg^{-1}$. According to Newton's second law, $\frac{F}{m}$ is the acceleration of free fall g therefore $a = g$. On the surface of the Earth $a = g = 9.81\,N\,kg^{-1}$.

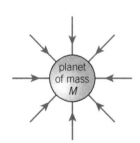

▲ **Figure 2** *A spherical mass M can be modelled as a point mass M*

18.2 Newton's law of gravitation

Newton's law of gravitation is a universal law that can be applied to all objects.

Newton's law of gravitation: Two point masses attract each other with a force that is directly proportional to the product of their masses and inversely proportional to the square of the separation.

Equation for Newton's law

According to Newton's law of gravitation

$$F \propto -\frac{Mm}{r^2}$$

where F is the gravitational force, M and m are the masses, and r is the separation. The minus sign means an attractive force F. The law can be written as an equation using the **gravitational constant** G ($6.67 \times 10^{-11}\,N\,kg^{-1}$) as follows:

$$F = -\frac{GMm}{r^2}$$

The gravitational field of a spherical object (e.g., planet) can be modelled as a **point mass** at its centre, see Figure 2. So just remember that r is the centre-to-centre separation, see Figure 3.

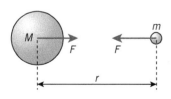

▲ **Figure 3** *The separation r is from centre-to-centre*

18.3 Gravitational field strength for a point mass

The **gravitational field strength** g depends on the distance r from the centre of a spherical mass. For the Earth, g is only equal to $9.81\,\text{N kg}^{-1}$ at the surface. Beyond its surface, g obeys an inverse square law with distance.

Radial field

A spherical object of mass m produces a radial field. The gravitational field strength g at a distance r from the centre of the mass can be determined as follows:

$$g = F \div m = -\frac{GMm}{r^2} \div m$$

Therefore

$$g = -\frac{GM}{r^2}$$

Figure 4 shows the variation of g with r for a point or spherical mass.

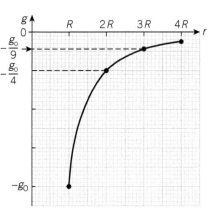

▲ **Figure 4** *Gravitational field strength g obeys an inverse square law*

> **Worked example: Mass of the Moon**
>
> The Moon has diameter 3480 km and surface gravitational field strength of $1.62\,\text{N kg}^{-1}$.
>
> Calculate the mass of the Moon.
>
> **Step 1:** Write down the quantities needed to calculate the mass.
>
> $r = 1740 \times 10^3\,\text{m} \quad g = 1.62\,\text{N kg}^{-1} \quad G = 6.67 \times 10^{-11}\,\text{N m}^2\,\text{kg}^{-2}$
>
> **Step 2:** Rearrange the equation for g and then substitute the values.
>
> $g = \dfrac{GM}{r^2}$ (magnitude only)
>
> $M = \dfrac{gr^2}{G} = \dfrac{1.62 \times (1740 \times 10^3)^2}{6.67 \times 10^{-11}} = 7.35 \times 10^{22}\,\text{kg}$ (3 s.f.)

> **Revision tip: Inverse square law**
>
> $g \propto \dfrac{1}{r^2}$ which is an inverse square law. Doubling the distance r from the centre of the object will decrease g by a factor of $2^2 = 4$.

Summary questions

1. The gravitational force on an object of mass 50 kg is 190 N.
   Calculate the gravitational field strength. *(1 mark)*
2. The gravitational field strength at a point is $3.0\,\text{N kg}^{-1}$.
   Calculate the gravitational force experienced by a space probe of mass 600 kg at this point. *(1 mark)*
3. Calculate the gravitational force between Mercury and the Sun. *(2 marks)*
   mass of Sun = $2.0 \times 10^{30}\,\text{kg}$
   mass of Mercury = $3.3 \times 10^{23}\,\text{kg}$
   separation = $5.8 \times 10^{10}\,\text{m}$
4. The gravitational force experienced by a rocket is F on the surface of the Earth.
   Calculate the force (in terms of F) on the rocket at a *height* of 3 Earth radii. *(3 marks)*
5. The mass of Uranus is $8.7 \times 10^{25}\,\text{kg}$ and its surface gravitational field strength is $10\,\text{N kg}^{-1}$.
   Calculate its radius. *(3 marks)*
6. Compared with the Earth, Jupiter has 320 times greater mass and is 11 times larger.
   Estimate the gravitational field strength on the surface of Jupiter. *(3 marks)*

18.4 Kepler's laws
18.5 Satellites

Specification reference: 5.4.3

18.4 Kepler's laws

The German mathematician and astronomer Johannes Kepler (1571–1630) published his three laws of planetary motion in the 17th century based on the naked-eye planetary observations of the Danish astronomer Tycho Brahe.

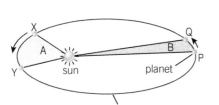

▲ Figure 1 *A planet travels from X to Y and from P to Q in the same time. According to the second law, area A = area B*

The three laws

First law: All planets move in elliptical orbits, with the Sun at one focus.

Second law: A line that connects a planet to the Sun sweeps out equal areas in equal times. (See Figure 1.)

Third law: The square of the orbital period of any planet is directly proportional to the cube of the mean distance from the Sun.

The third law may be written as

$$T^2 \propto r^3 \text{ or } \frac{T^2}{r^3} = K$$

where K is a constant for the planets orbiting the Sun.

Although the third law originated for the planets in our Solar System, it can be applied to any system where objects orbit round a central mass. For example, you can apply this law to the satellites orbiting the Earth, to the moons of Jupiter, to planets orbiting other stars in our galaxy, and so on.

Modelling planetary motion

Figure 2 shows a planet of mass m moving at a speed v in a circular orbit of radius r round the Sun. The orbital period of the planet is T. The mass of the Sun is M. You can use Newton's law of gravitation and ideas from circular motion to show the validity of Kepler's third law.

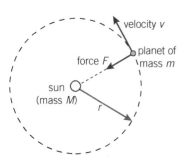

▲ Figure 2 *A planet orbiting the Sun*

The centripetal force on the planet is provided by the gravitational force. Therefore

$$F = \frac{mv^2}{r}$$

$$\frac{GMm}{r^2} = \frac{mv^2}{r}$$

$$v^2 = \frac{GM}{r}$$

$$\left(\frac{2\pi r}{T}\right)^2 = \frac{GM}{r}$$

$$T^2 = \left(\frac{4\pi^2}{GM}\right) r^3$$

For the planets in the Solar System, GM is constant therefore

$$T^2 \propto r^3$$

which is Kepler's third law.

Revision tip: Graphs

1. A graph of T^2 against r^3 will be a straight line of gradient $\frac{4\pi^2}{GM}$ and passing through the origin.

2. A graph of $\lg T$ against $\lg r$ should be linear, with a gradient of 1.5.

Synoptic link

Centripetal forces were covered in Topic 16.3, Exploring centripetal forces.

18.5 Satellites

Artificial **satellites** are used for space exploration, exploration of the Earth, communications, and so on. The physics of any satellite can be analysed using Newton's law of gravitation and equations for circular motion.

Geostationary

A satellite in a geostationary orbit:

- has an orbital period of 1 day round the Earth
- travels in the same direction as the rotation of the Earth
- has its orbit in the equatorial plane of the Earth.

The distance r of a satellite in a geostationary orbit can be calculated using the equation for Kepler's third law.

$$T^2 = \left(\frac{4\pi^2}{GM}\right)r^3$$

The radius of a geostationary orbit is about 6.6 Earth radii from the centre of the Earth.

> **Worked example: Moons of Jupiter**
>
> Jupiter's moon Io is 420 Mm from the centre of Jupiter and has a period of 1.8 days. Observations of Europa, another moon of Jupiter, shows it to have a period of 3.6 days. Calculate the radius of orbit of Europa.
>
> **Step 1:** Write down the known and unknown quantities for Kepler's third law.
>
> Io: $T = 1.8$ days and $r = 420$ Mm   Europa: $T = 3.6$ days and $r = ?$
>
> **Step 2:** Use Kepler's third law to calculate r.
>
> $$\frac{T^2}{r^3} = \text{constant}$$
>
> $$\frac{1.8^2}{420^3} = \frac{3.6^2}{r^3}$$
>
> $$r = \sqrt[3]{\frac{420^3 \times 3.6^2}{1.8^2}} = 670 \text{ Mm (2 s.f.)}$$
>
> *You can work in any consistent units*

Summary questions

Data for the Earth: mass = 6.0×10^{24} kg   radius = 6400 km

1. State Kepler's third law. *(1 mark)*
2. State the orbital period of a satellite in geostationary orbit. *(1 mark)*
3. Calculate the speed of a satellite orbiting at a distance of 3000 km *above* the surface of the Earth. *(4 marks)*
4. Calculate the orbital period of the satellite in **Q3**. *(3 marks)*
5. For planets in the Solar System, explain why a graph of lg$T$ against lg$r$ has gradient 1.5. *(3 marks)*
6. Show that the radius of a geostationary orbit for the Earth is about 6.6 Earth radii. *(3 marks)*

18.6 Gravitational potential
18.7 Gravitational potential energy

Specification reference: 5.4.4

18.6 Gravitational potential

You need the idea of gravitational potential in order to understand the gain and loss in the gravitational potential energy of objects such as rockets and planets.

Gravitational potential V_g

The **gravitational potential** V_g at a point in a gravitational field is defined as the work done per unit mass to move an object to that point from infinity.

For a spherical object, such as a planet or star, the gravitational potential V_g is directly proportional to the mass m of the object and inversely proportional to the distance r from the centre of the object. The equation for V_g is

$$V_g = -\frac{GM}{r}$$

where G is the gravitational constant $6.67 \times 10^{-11}\,\text{N}\,\text{m}^2\,\text{kg}^{-2}$.

- V_g has unit $\text{J}\,\text{kg}^{-1}$.
- V_g is defined to be zero at infinity.
- The negative sign signifies that the gravitational force is attractive.

At the surface of the Earth, V_g is about $-63\,\text{MJ}\,\text{kg}^{-1}$. It will take a minimum energy of 63 MJ to remove 1 kg mass all the way to infinity. The 1 kg mass will completely escape the gravitational influence of the Earth if given an energy slightly greater than 63 MJ.

Figure 1 shows the variation of V_g with distance r.

Revision tip: Graphs
Gravitational potential V_g is always negative.

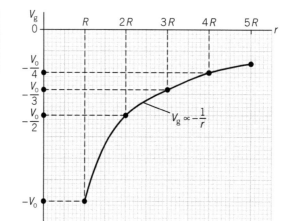

▲ **Figure 1** V_g obeys an inverse law with distance r

Equipotentials: Extension work

At a specific distance r from the centre of a spherical mass, V_g has a specific negative value. An equipotential is a line of equal gravitational potential. The equipotentials for a radial field are concentric circles. Figure 2 shows the equipotentials of an imaginary planet.

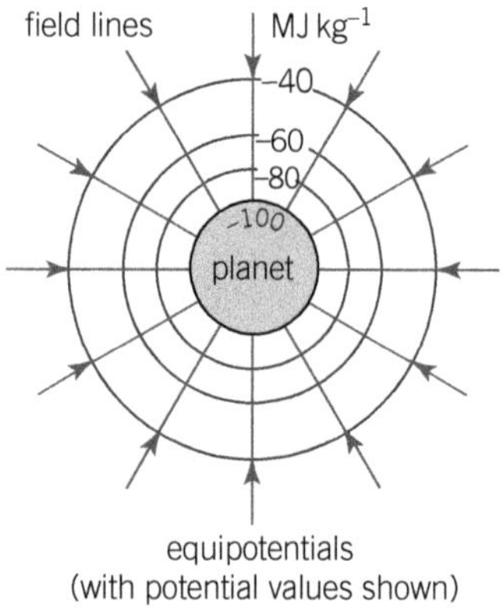

18.7 Gravitational potential energy

The **gravitational potential energy** E of any object with mass m within a gravitational field is defined as the work done to move the mass from infinity to a point in a gravitational field.

▲ **Figure 2** Equipotentials for a planet. The field lines are at right angles to the equipotentials

Gravitational fields

From the definition of gravitational potential energy, we have

$$E = mV_g$$

$$E = -\frac{GMm}{r}$$

A comet, in an elliptical orbit, has both kinetic energy and gravitational potential energy. Its total energy in its orbit remains constant. Its kinetic energy increases as it gets closer to the Sun and its potential energy decreases.

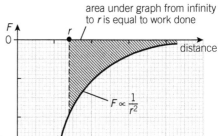

▲ **Figure 3** *The area under the force–distance graph is work done*

Understanding gravitational potential energy

Figure 3 shows the force F against distance graph for a spherical mass (planet). The area under the force–distance graph is equal to work done. The area under the graph from infinity to a distance r is equal to the gravitational potential energy E.

Quite often, we are interested in changes in potential and potential energy.

- change in gravitational potential ΔV_g = final V_g – initial V_g
- change in gravitational potential energy ΔE = final E – initial E

Escape velocity

Escape velocity of a projectile is the minimum velocity it must have to completely escape from the gravitational influence of a planet.

You can determine the escape velocity v from a planet of mass M and radius R as follows:

initial total energy = KE + PE = $\frac{1}{2}mv^2 - \frac{GMm}{R}$

final total energy of projectile = 0

Therefore

$$\frac{1}{2}mv^2 = \frac{GMm}{R}$$

$$v = \sqrt{\frac{2GM}{R}}$$

🖩 Worked example: Escaping from the Moon

The mass of the Moon is 7.4×10^{22} kg and it has radius 1700 km.

Calculate the escape velocity from the Moon.

Step 1: Write down the information given.

$M = 7.4 \times 10^{22}$ kg $R = 1700 \times 10^3$ m $G = 6.67 \times 10^{-11}$ N m² kg⁻²

Step 2: Calculate the escape velocity v.

$$v = \sqrt{\frac{2GM}{R}} = \sqrt{\frac{2 \times 6.67 \times 10^{11} \times 7.4 \times 10^{22}}{1700 \times 10^3}}$$

$v = 2.4$ km s⁻¹ (2 s.f.)

Summary questions

Data for the Earth: mass = 6.0×10^{24} kg, radius = 6400 km

1. State the maximum value for gravitational potential. *(1 mark)*

2. The gravitational potential on the surface of Venus is -54 MJ kg⁻¹. What is the energy required to remove a 1 kg mass from the surface of Venus to infinity? *(1 mark)*

3. The radius of Venus is 6.1 Mm. Use the information in **Q2** to determine its mass. *(2 marks)*

4. Calculate the gravitational potential at the surface of the Earth. *(2 marks)*

5. A rocket of mass 1200 kg is moved from the surface of the Earth to an orbit at a distance of 9400 km from the centre of Earth. Calculate the change in gravitational potential. *(3 marks)*

6. Calculate the change in gravitational potential energy of the rocket in **Q5**. *(2 marks)*

Chapter 18 Practice questions

1. A satellite is in a geostationary orbit.
   What is the period of rotation of this satellite around the Earth?
   A 3.60×10^3 s
   B 4.32×10^4 s
   C 8.64×10^4 s
   D 3.16×10^7 s (1 mark)

2. A planet has a radius R. The gravitational field strength on the surface of the planet is g_0.
   What is the distance from the *surface* of the planet where the gravitational field strength is $\frac{g_0}{9}$?
   A $2R$
   B $3R$
   C $8R$
   D $9R$ (1 mark)

3. Astronomers have recently discovered a distant star to have orbiting planets. The period of a planet is T and its distance from the star is r.
   Which graph will *not* produce a straight line graph?
   A T^2 against r^3
   B T against $r^{1.5}$
   C $\lg(T)$ against r
   D $\lg(T)$ against $\lg(r)$ (1 mark)

4. a A small object is falling towards the centre of a planet. Explain why its acceleration of free fall at a point is the same as the gravitational field strength at that point. (2 marks)
   b The Earth has a mean radius of 6400 km. The table below shows the magnitude of the gravitational field strength g and distance r from the centre of the Earth.

r / km	6400	8000	9600
g / N kg^{-1}	9.8	6.3	4.4

   i Use the table to confirm the relationship between g and r. (2 marks)
   ii Calculate the mass of the Earth. (3 marks)
   iii Calculate the mean density of the Earth. (2 marks)

5. A planet of mass m is orbiting in a circular path at a distance r from the centre of the star. The mass of the star is M.
   a Derive an equation for the speed v of the planet in its orbit. (3 marks)
   b Neptune is about 30 times further from the Sun than the Earth.
   Calculate the ratio $\dfrac{\text{speed of Earth}}{\text{speed of Neptune}}$. (2 marks)

6. Figure 1 shows the Moon. The value of the gravitational potential at its surface is -2.8 MJ kg^{-1}. The radius of the Moon is R.
   a How much energy would it take to completely remove a 1.0 kg mass from the surface of the Moon? (1 mark)
   b Determine the gravitational potentials at points A and B. (3 marks)
   c Calculate the energy required to move a rocket of mass 3500 kg from A to B. (2 marks)

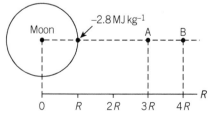

▲ Figure 1

Chapter 19 Stars

In this chapter you will learn about ...

- [ ] Stellar formation
- [ ] Life cycles of stars
- [ ] Hertzsprung–Russell (HR) diagrams
- [ ] Energy levels
- [ ] Line spectra
- [ ] Emission, absorption, and continuous spectra
- [ ] Diffraction grating
- [ ] Stellar luminosity
- [ ] Wien's displacement law
- [ ] Stefan's law

19 STARS
19.1 Objects in the Universe
19.2 The life cycle of stars
19.3 Hertzsprung–Russell diagram

Specification reference: 5.5.1

19.1 Objects in the Universe

The **Universe** is immense in size. It has about 10^{11} galaxies – each galaxy has about 10^{11} stars. Our local star is the Sun – it has orbiting planets, asteroids, comets, and dust.

Stellar formation

Stars are formed from large clouds of dust and gas (mainly hydrogen) called **nebulae**.

- The tiny gravitational force between dust and gas slowly brings some parts of the cloud together (gravitational collapse).
- The gravitational potential energy of the cloud decreases as its kinetic energy and temperature increases.
- *Fusion* of hydrogen nuclei into helium nuclei occurs when the temperature of the cloud is about 10^7 K. The energy released from the fusion reactions further increases the temperature. A hot ball of gas (star) is formed.
- The gravitational forces compress the star. The **radiation pressure** from the photons emitted during fusion and the **gas pressure** of the star push outwards. A stable star is formed when the force from the radiation and gas pressure is balanced by the gravitational force.

> **Synoptic link**
>
> There is more detail on nuclear fusion in Topic 26.4, Nuclear fusion.

19.2 The life cycle of stars

The rate of fusion subsides when most of the hydrogen is depleted in a star. The core of a star is made of rings of elements with iron at the centre. The outermost layers of the star has helium and a very small amount of hydrogen. The fusion of helium makes outermost layers of the star expand into a **red giant**.

Low-mass stars

Figure 1 shows the evolution of a star with mass between $0.5\,M_\odot$ and $10\,M_\odot$.

$M_\odot$ is the solar mass 2.0×10^{30} kg.

- Eventually most of the layers of the red giant around the core drift away into space as a **planetary nebula.**
- The hot core (30 000 K) left behind is called a **white dwarf.**
- The white dwarf is very dense. There is no fusion but it slowly leaks away photons created in its earlier evolution. It is prevented from gravitational collapse by **electron degeneracy pressure.** The maximum mass of a white dwarf is $1.44 M_\odot$. This is called the **Chandrasekhar limit**.

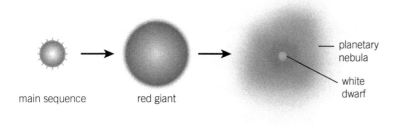

▲ **Figure 1** *Evolution of a low-mass star, for example the Sun*

Massive stars

Figure 2 shows the evolution of a star of mass greater than about $10M_\odot$.

- When no further fusion reactions occur, the layers in the core suddenly implode under gravitational forces and bounce off the solid core, leading to a shockwave that ejects core material into space. This event is called a (type II) **supernova**.
- The remnant core is a **neutron star** when its mass is greater than the Chandrasekhar limit. The core has tightly packed neutrons and extremely dense (~ 10^{17} kg m^{-3}).
- The remnant core is a **black hole** when its mass is greater than about $3M_\odot$. Black holes could be a singularity and have infinite density. Light cannot escape from a black hole.

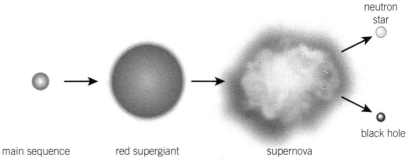

▲ **Figure 2** *Evolution of a massive star*

19.3 Hertzsprung–Russell diagram

The **Hertzsprung–Russell (HR) diagram** is a graph of stars in a galaxy, or a star cluster. **Luminosity** is plotted on the *y*-axis and the surface temperature of the stars on the *x*-axis. The temperature axis has temperature increasing from right to left (which is very odd).

The luminosity L of a star is the total radiant power of the star.

HR diagram

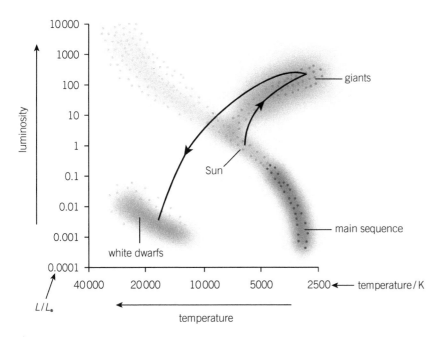

▲ **Figure 3** *HR diagram showing the likely evolution of the Sun. $L_\odot$ is the luminosity of the Sun*

Figure 3 shows the HR diagram for the stars in a galaxy. The bulk of the stars lie on the **main sequence** band. The white dwarfs and red giants form their own distinct regions.

Summary questions

1. State the Chandrasekhar limit. *(1 mark)*
2. Calculate the mass of a $10\,M_\odot$ star. *(1 mark)*
3. State the quantity that governs the fate of the remnant core of a massive star. *(1 mark)*
4. Compare the white dwarfs and the red giants on the HR diagram. *(2 marks)*
5. A particular neutron star has mass 4.0×10^{30} kg and a radius of only 11 km. Calculate the mass of a 1 mm^3 'grain' of neutron star matter. *(3 marks)*
6. Use you knowledge of escape velocity to estimate the maximum radius of a black hole of mass $5\,M_\odot$ *(3 marks)*

19.4 Energy levels in atoms
19.5 Spectra

Specification reference: 5.5.2

19.4 Energy levels in atoms

The electrons moving inside the atoms behave as de Broglie waves and produce stationary waves because they are trapped within the atoms. As a result, the electrons are only allowed a discrete set of energy values – their energy is quantised.

An **energy level** is one of the discrete set of energies a bound electron can have.

Quantised energy levels

Figure 1 shows a typical energy level diagram for the electrons in gas atoms. The vertical scale has the energy of the electron. The horizontal lines indicate the permitted energy of the electrons.

- An electron cannot exist between energy levels.
- An electron can move from one energy level to another.
- All energy levels have negative values. This simply means that external energy is required to pull away the electrons from the attractive electrostatic forces of the positive nuclei.
- The **ground state** is the most negative energy level.

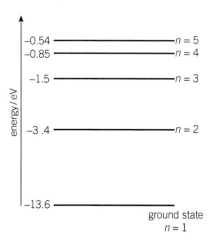

▲ **Figure 1** *Energy levels for the hydrogen atom. The levels are labelled with the principal quantum number n*

Emission lines

What happens when an electron makes a transition from a higher energy level to a lower energy level? Energy must be conserved. The transition of the electron produces a photon of a unique wavelength, see Figure 2. The energy of the photon is equal to the difference between the energy levels ΔE. Therefore

$$hf = \Delta E \quad \text{or} \quad \frac{hc}{\lambda} = \Delta E$$

where h is the Planck constant, f is the frequency of the electromagnetic radiation, and λ is the wavelength. An emission spectrum has sharp and bright lines, each corresponding to the unique wavelength originating from electron transitions between a pair of energy levels.

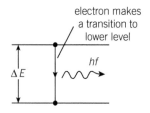

▲ **Figure 2** *Understanding emission spectrum*

19.5 Spectra

There are three types of spectra: **continuous**, **emission**, and **absorption**. The origin of **emission line spectrum** has already been outlined above.

Continuous and absorption spectra

The atoms of a heated solid metal, such as a lamp filament, produce a continuous spectrum. This type of spectrum will have all the wavelengths of visible light from blue to red – effectively all the rainbow colours.

Absorption line spectrum is characterised by a series of dark lines against the background of a continuous spectrum. An absorption spectrum is produced when the electrons within cooler gas atoms absorb photons. Figure 3 shows a photon being absorbed by an electron. The electron will make a transition (a jump) to a higher energy level only when the photon has the right amount of energy ΔE. The photon disappears. After some time, the electron will make a transition to the lower energy level and re-emit the photon. The photons are emitted in random directions, so the intensity in the original direction is greatly reduced.

> **Synoptic link**
> You studied photons in Topic 13.1, The photon model, and Topic 13.2, The photoelectric effect.

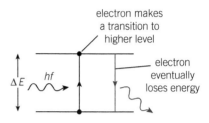

▲ **Figure 3** *Understanding absorption spectrum*

> **Worked example: Emission line**
>
> An electron makes a transition from −5.0 eV level to −6.1 eV level.
>
> Calculate the wavelength of the photon.
>
> **Step 1:** Calculate the energy loss of the electron in J.
>
> energy loss = 6.1 − 5.0 = 1.1 eV
>
> energy loss = $1.1 \times 1.6 \times 10^{-19} = 1.76 \times 10^{-19}$ J
>
> **Step 2:** Calculate the wavelength.
>
> $$\frac{hc}{\lambda} = \Delta E$$
>
> $$\lambda = \frac{6.63 \times 10^{-34} \times 3.00 \times 10^{8}}{1.76 \times 10^{-19}} = 1.13 \times 10^{-6} \text{ m}$$
>
> wavelength = 1.1×10^{-6} m (2 s.f.)

> **Revision tip**
> 1 eV = 1.60×10^{-19} J

Summary questions

1. Explain what is meant by an energy level. *(1 mark)*
2. An electron makes a jump from the −2.0 eV energy level to the −6.0 eV energy level.
   Calculate the energy in eV of the emitted photon. *(1 mark)*
3. An electron makes a transition from the −10 eV level to the −2.0 eV level. Explain whether this involves the emission or the absorption of a photon. *(1 mark)*
4. Calculate the wavelength of the photon emitted in Q2. *(3 marks)*
5. Determine the maximum number of possible emission spectral lines from 4 energy levels. *(2 marks)*
6. An electron is in the −12.2 eV level. It absorbs a photon of wavelength 400 nm.
   Calculate the value of the new energy level in eV. *(4 marks)*

19.6 Analysing starlight
19.7 Stellar luminosity

Specification reference: 5.5.2

19.6 Analysing starlight

A **diffraction grating** consists of a large number of regularly spaced lines on a glass or plastic slide. Each line behaves like a narrow slit. A parallel beam of monochromatic light directed normally at the grating will be diffracted at each slit (line) and the diffracted light will then interfere beyond the slits. Bright light is only transmitted by the grating in certain directions.

Figure 1 shows the maxima (bright light) and the numbering (orders) of these maxima.

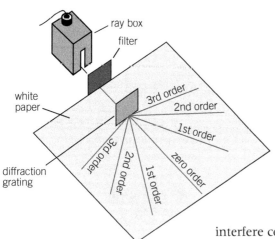

▲ **Figure 1** *A diffraction grating produces bright light in certain directions*

Transmission diffraction grating

Figure 2 shows two rays of diffracted light from adjacent slits of the grating. At the angle θ, the light from the slits P and Q interfere constructively. The phase difference must be zero and the path difference must be a whole number of wavelengths. Therefore

$$QY = n\lambda = d\sin\theta$$

This gives us the grating equation

$$d\sin\theta = n\lambda$$

where d is the separation between adjacent slits (known as **grating spacing**), λ is the wavelength of the monochromatic light, and n is the order.

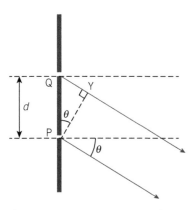

▲ **Figure 2** *Formation of the nth order maxima*

> ### 🖩 Worked example: Stellar spectral line
>
> The light from a star is analysed using a diffraction grating with 500 lines per mm.
>
> A bright spectral line is observed at an angle of 28° in the second-order spectrum.
>
> Calculate the wavelength of this spectral line in nm.
>
> **Step 1:** Calculate the grating spacing d.
>
> $$d = \frac{1.0 \times 10^{-3}}{500} = 2.0 \times 10^{-6} \, \text{m}$$
>
> **Step 2:** Write down the quantities and then use the grating equation to calculate λ.
>
> $n = 2 \quad d = 2.0 \times 10^{-6}\,\text{m} \quad \theta = 24° \quad \lambda = ?$
>
> $d \sin\theta = n\lambda$
>
> $$\lambda = \frac{2.0 \times 10^{-6} \times \sin 28°}{2} = 4.7 \times 10^{-7}\,\text{m}$$
>
> wavelength = 470 nm (2 s.f.) $\quad$ (1 nm = 10^{-9} m)

Synoptic link

You studied path and phase differences in Topic 12.1, Superposition of waves, 12.2, Interference, and 12.3, The Young double-slit experiment.

19.7 Stellar luminosity

Stars and many other hot objects can be modelled as a **black body**. Figure 3 shows a typical graph of intensity against wavelength of electromagnetic waves from a black body. The shape of the graph is unique to the surface temperature of the body. The wavelength λ_{max} at the peak of the graph is related to the thermodynamic temperature T of the surface of the body.

Wien's displacement law

Wien's displacement law: the black body radiation curve peaks at a wavelength that is inversely proportional to the thermodynamic temperature of the body.

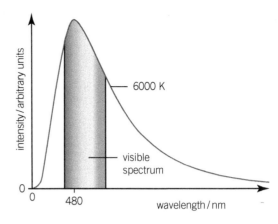

▲ **Figure 3** *Intensity–wavelength graph for a black body at 6000 K*

Therefore
$$\lambda_{max} \propto \frac{1}{T} \quad \text{or} \quad \lambda_{max} T = \text{constant}$$

The value of the constant is 2.9×10^{-3} m K.

Stefan's law

The Stefan–Boltzmann law (also known as Stefan's law): the total power radiated per unit surface area of a black body is directly proportional to the fourth power of its thermodynamic temperature.

The luminosity L of a star is the total power it emits. For a star of radius r, the total surface area is $4\pi r^2$, and the luminosity is given by the relationship

$$L \propto 4\pi r^2 T^4$$

With the **Stefan constant** σ (5.67×10^{-8} W m^{-2} K^{-4}), the relationship above can be written as an equation.

$$L = 4\pi r^2 \sigma T^4 \quad \text{(Note: } L \propto r^2 \text{ and } L \propto T^4\text{)}$$

How big is a star?

The radiation from our Sun peaks at a wavelength of 5.0×10^{-7} m and the surface temperature of the Sun is about 5800 K. The temperature of a star can be determined using Wien's law. If its luminosity is known, then the equation $L = 4\pi r^2 \sigma T^4$ can be used to determine its radius r.

Summary questions

1. Calculate the grating spacing in metre (m) for a grating with 80 lines per mm. *(1 mark)*
2. State how the luminosity of a star depends on its surface temperature. *(1 mark)*
3. Monochromatic light of wavelength 6.4×10^{-7} m is incident normally at a grating with 800 lines per mm.
   Calculate the angle θ for the first-order maxima. *(3 marks)*
4. The surface temperature of the Sun is 5800 K and its peak wavelength is 500 nm.
   Calculate the peak wavelength for Rigel which has a temperature of 12 000 K. *(2 marks)*
5. Explain why red giants are very luminous compared with similar temperature main sequence stars. *(2 marks)*
6. Rigel has a radius 79 times that of the Sun.
   Use information in Q4 to calculate the luminosity of Rigel in term of the luminosity $L_\odot$ of the Sun. *(4 marks)*

Chapter 19 Practice questions

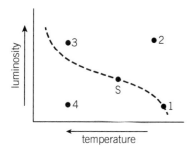

▲ Figure 1

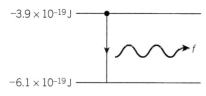

▲ Figure 2

1. Figure 1 shows an incomplete Hertzsprung–Russell (HR) diagram. The position of the Sun is marked by the letter S. The numbers 1, 2, 3, and 4 are positions on this HR diagram.

   What is the likely evolution of the Sun?

   A S → 2 → 1
   B S → 3 → 4
   C S → 4 → 1
   D S → 2 → 4 *(1 mark)*

2. Figure 2 shows an electron making a transition between two energy levels.
   What is the frequency f of the emitted photon?

   A 3.3×10^{14} Hz
   B 5.9×10^{14} Hz
   C 9.2×10^{14} Hz
   D 1.5×10^{15} Hz *(1 mark)*

3. Star X has radius R, surface temperature T, and luminosity L. Star Y has radius $2R$ and surface temperature $2T$.
   What is the luminosity of star Y in terms of L?

   A $4L$
   B $8L$
   C $32L$
   D $64L$ *(1 mark)*

4. a Describe some of the characteristics of a white dwarf. *(2 marks)*

   b i Sketch a graph to show the variation of intensity of a star with wavelength. *(1 mark)*

      ii State the relationship between the wavelength at maximum intensity and the surface temperature of the star. *(1 mark)*

   c The surface temperature of the Sun is 5800 K and its luminosity is 3.8×10^{26} W.

      Sirius-B is a white dwarf with a surface temperature of about 25 000 K and a radius that is about 120 times smaller than the radius of the Sun. Estimate the luminosity of Sirius-B. *(3 marks)*

5. The visible spectrum of a distant star is analysed using a diffraction grating with 800 lines mm^{-1}.

   a Explain how elements can be identified in the atmosphere of stars by analysing the spectrum of the light they emit. *(2 marks)*

   b The diffraction grating is used to view the hydrogen emission line of wavelength 656 nm.

      Calculate the maximum number of orders that can be observed using light of this wavelength. *(3 marks)*

   c Calculate the angular separation between two emission lines observed in the first-order spectrum that have wavelengths 589 nm and 587 nm. *(3 marks)*

6. a Explain what is meant by the luminosity of a star. *(1 mark)*

   b A star has radius 8.5×10^5 km and a surface temperature of 5800 K. It is 3.9×10^{16} m from the Earth. Calculate:

      i the luminosity of the star; *(2 marks)*

      ii the intensity of the light from this star at the Earth. *(2 marks)*

   c Suggest why the actual intensity of the light from the star in b at the Earth will be less than your answer to b(ii). *(1 mark)*

Chapter 20 Cosmology (The Big Bang)

In this chapter you will learn about ...

- ☐ Astronomical unit (AU)
- ☐ Light year (ly)
- ☐ Parsec (pc)
- ☐ Doppler effect
- ☐ Doppler equation
- ☐ Hubble's law
- ☐ Cosmological principle
- ☐ The Big Bang theory
- ☐ The evolution of the Universe

20 COSMOLOGY (THE BIG BANG)
20.1 Astronomical distances
20.2 The Doppler effect

Specification reference: 5.5.3

20.1 Astronomical distances

In order to visualise the scale of space and separation between objects in the Universe, astronomers use astronomical unit, light-year and parsec to measure distances.

Astronomical unit (AU)

The **astronomical unit** is the average distance between the Earth and the Sun.

$$1 \text{ AU} = 1.50 \times 10^{11} \text{ m}$$

The AU is used to measure the distances within our own Solar System. For example, the distance of Saturn is 9.58 AU from the Sun.

Light-year (ly)

The light-year is the distance travelled by light in a vacuum in a time of one year.

$$1 \text{ ly} = \text{speed of light} \times \text{time}$$
$$1 \text{ ly} = 2.9979 \times 10^{8} \times (365.25 \times 24 \times 3600)$$
$$1 \text{ ly} = 9.46 \times 10^{15} \text{ m} \approx 9.5 \times 10^{15} \text{ m}$$

The light-year is used to measure distance between stars and galaxies. For example, the bright star Sirius is 8.60 ly from the Sun.

Parsec (pc)

The distance between stars and galaxies can also be measured in parsec (which comes from the abbreviation **par**allax of one **sec**ond of arc). In order to understand this unit, you need to first understand **stellar parallax** and angles measured in seconds of arc (or just arcseconds).

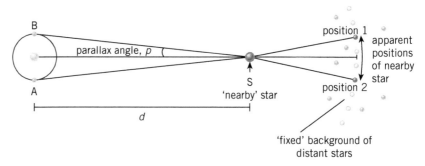

▲ **Figure 1** Stellar parallax

Stellar parallax is the apparent shift in the position of a nearby star when viewed against the background of very distant stars.

Figure 1 shows a nearby star observed from position A and then six months later from position B. The **parallax angle** p (also just known as parallax) is the angle subtended by a length of 1 AU at the position of the star. The parallax angle p is always significantly smaller than 1 degree. Very small angles are measured in seconds of arc – there are 3600 seconds of arc in 1°.

Cosmology (The Big Bang)

The parsec is the distance at which one astronomical unit (AU) subtends an angle of one arcsecond (see Figure 2).

Therefore

$$\tan\left(\tfrac{1}{3600}^\circ\right) = \frac{1\,\text{AU}}{1\,\text{pc}} = \frac{1.50 \times 10^{11}}{1\,\text{pc}} \quad \text{(See Figure 2)}$$

$$1\,\text{pc} = \frac{1.50 \times 10^{11}}{\tan\left(\tfrac{1}{3600}^\circ\right)} = 3.09 \times 10^{16}\,\text{m} \approx 3.1 \times 10^{16}\,\text{m}$$

The parallax angle p in arcsconds and the distance d in pc of a star are related by the equation $d = \dfrac{1}{p}$.

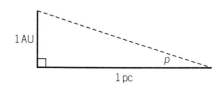

▲ Figure 2

20.2 The Doppler effect

Doppler effect is the change in frequency or wavelength of a wave when there is relative velocity between a source and the observer.

You can observe Doppler effect with all waves, including sound and light. Doppler effect can be used to determine the relative speed of stars in our galaxy and of other galaxies too.

Light from moving sources

- If a source of light is moving towards the Earth, the entire spectrum observed from the source is shifted to shorter wavelengths. This is known as **blue-shift**.
- If a source of light is receding away the Earth, the entire spectrum observed from the source is shifted to longer wavelengths. This is known as **red-shift**.

The light from distant galaxies shows red-shift. All distant galaxies are receding away from each other. The light from Andromeda, our nearest galaxy, shows blue-shift. It is very slowly moving towards our galaxy

Doppler equation

The speed v of a star or a galaxy can be determined using the **Doppler equation** below:

$$\frac{\Delta \lambda}{\lambda} \approx \frac{v}{c}$$

where $\Delta \lambda$ is the change in wavelength, λ is the wavelength measured in the laboratory, and c is the speed of light in a vacuum. This equation can only be used when v is much smaller than c. You can also use the following equation:

$$\frac{\Delta f}{f} \approx \frac{v}{c}$$

where Δf is the change in frequency and f is the frequency.

> ### Worked example: Speed of a distant galaxy
>
> In the laboratory an absorption line of hydrogen is observed at a wavelength of 656.4 nm. The same spectral light from the distant galaxy is observed at 664.7 nm. Calculate the speed of this receding galaxy.
>
> **Step 1:** Calculate the change in the wavelength.
>
> $\Delta \lambda = 664.7 - 656.4 = 8.3\,\text{nm}$
>
> **Step 2:** Substitute values into the Doppler equation and calculate the speed v.
>
> $$\frac{\Delta \lambda}{\lambda} \approx \frac{v}{c}$$
>
> $$v \approx \frac{c \times \Delta \lambda}{\lambda} \approx \frac{3.0 \times 10^8 \times 8.3}{656.4} = 3.8 \times 10^6\,\text{ms}^{-1}\,(2\,\text{s.f.})$$

Summary questions

1. Change the following distances into metres (m):
   a 5.2 AU   b 1.6 ly
   c 2600 ly   d 95 pc
   (4 marks)

2. Convert 0.78 arcseconds into degrees. *(1 mark)*

3. The centre of our galaxy is about 8.0 kpc from the Earth. Calculate this distance in metres (m) and the time it takes light to travel this distance. *(3 marks)*

4. The star Sirius is travelling towards us at a speed of 7.6 km s^{-1}. Calculate the percentage change in the wavelength of a specific spectral line. *(2 marks)*

5. Sirius is 8.6 ly away from us. Calculate its parallax angle in arcseconds. *(3 marks)*

6. The wavelength of a specific spectral line in the laboratory is 119.5 nm. The same spectral line is observed in the spectrum of a star moving away from us at a speed of 5.3×10^6 m s^{-1}. Calculate this observed wavelength. *(3 marks)*

45

20.3 Hubble's law
20.4 The Big Bang theory
20.5 Evolution of the Universe

Specification reference: 5.5.3

20.3 Hubble's law

All distant galaxies are receding from each other. This provides the evidence for the **Big Bang** theory of the **expanding Universe**. The galaxies are moving apart because the whole fabric of space has been, and is still, expanding.

Hubble constant

Hubble's law: The recession speed v of a galaxy is directly proportional to its distance d from us.

Therefore

$$v \propto d \quad \text{or} \quad v = H_0 d$$

where H_0 is the **Hubble constant**.

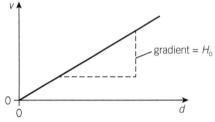

▲ **Figure 1** *A graph of v against d for galaxies*

- A graph of v against d is a straight-line graph through the origin. See Figure 1.
- The gradient of the line is H_0.
- The SI unit for the Hubble constant is s^{-1}, but it is often quoted in km s^{-1} Mpc^{-1} (kilometres per second per unit mega parsec). H_0 is about 2.2×10^{-18} s^{-1} or 70 km s^{-1} Mpc^{-1}.
- The age of the Universe is given by the equation: age of Universe = H_0^{-1}.

Cosmological principle

The cosmological principle: When viewed on a large enough scale, the Universe is **homogeneous** and **isotropic**, and the laws of physics are universal.

The principle is simply stating that the Universe is the same for all observers in the Universe.

- The laws of physics are the same for all observers in the Universe.
- Homogeneous means that matter is distributed uniformly across the Universe.
- Isotropic means that the Universe looks the same in all directions to every observer.

20.4 The Big Bang theory

The Big Bang theory is the standard model that describes the origin of the Universe and its subsequent large-scale evolution.

- The Universe expanded from a singularity some 13.7 billion years ago.
- The Universe was extremely hot in the early stages.
- The expansion of the Universe over billions of years led to cooling to a temperature of 2.7 K.
- Treating the Universe as a black body at 2.7 K, the peak wavelength is about 1 mm, which is in the microwave region of the electromagnetic spectrum.
- There is alternative explanation for the microwave radiation – the early hot Universe mainly had energetic short-wavelength photons. The expansion of space stretched out the wavelength of these primordial photons, so they are now observed in the microwave region of the spectrum.

Cosmology (The Big Bang)

20.5 Evolution of the Universe

Figure 2 shows the composition of the Universe – matter is a very small contributor.

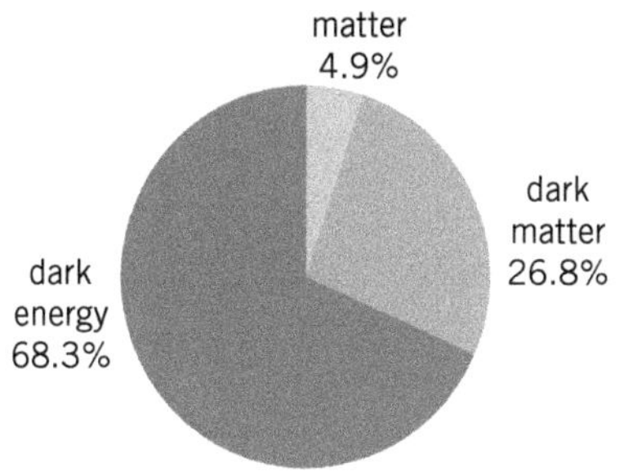

Evolution

time ↓

Space-time expansion began 13.7 billion years ago from a singularity.
↓
Expansion of the hot Universe led to cooling.
↓
Universe has high-energy photons, quarks, and leptons.
↓
Hadrons (protons and neutrons) formed.
↓
Fusion produces primordial helium – about 25% of the matter.
↓
Electrons combine with nuclei to form atoms.
↓
Stars formed and eventually galaxies.
↓

▲ **Figure 2** *Composition of the Universe*

The temperature of the Universe is now 2.7 K and we observe the same intensity of **microwave background radiation** in all directions (isotropic).

The receding galaxies, the existence of primordial helium in the very young galaxies, the temperature of 2.7 K, and the microwave background radiation all provide strong support for the Big Bang model of the Universe.

Worked example: How old?

The galaxy at a distance of 120 Mpc has a recession speed of 8000 km s^{-1}.

Determine an approximate age of the Universe based on this information.

Step 1: Calculate the Hubble constant in s^{-1}.

$$H_0 = \frac{v}{d} = \frac{8000 \times 10^3}{120 \times 10^6 \times 3.1 \times 10^{16}} = 2.15 \times 10^{-18}\,\text{s}^{-1}$$

(**Note:** 1 km s^{-1} = 10^3 m s^{-1} and 1 Mpc = $10^6 \times 3.1 \times 10^{16}$ m)

Step 2: Determine the age of the Universe.

$$\text{age} \approx H_0^{-1} = (2.15 \times 10^{-18})^{-1} = 4.65 \times 10^{17}\,\text{s}$$

$$\text{age} \approx 4.7 \times 10^{17}\,\text{s} \text{ (15 billion years) (2 s.f.)}$$

Summary questions

Use $H_0 = 70$ km s^{-1} Mpc^{-1} where required.

1. What is the current temperature of the Universe? *(1 mark)*
2. State the cosmological principle. *(1 mark)*

3. Calculate the speed in km s^{-1} of a galaxy at a distance of 200 Mpc. *(2 marks)*
4. Calculate the distance in Mpc of a galaxy receding at a speed of 5200 km s^{-1}. *(2 marks)*

5. Change 70 km s^{-1} Mpc^{-1} to s^{-1}. *(3 marks)*
6. Use Wien's displacement law and the information provided in Topic 19.7, Stellar luminosity, to show that the Universe is saturated with microwaves. *(3 marks)*

Chapter 20 Practice questions

1 When viewed on a large enough scale, the Universe is homogeneous and isotropic, and the laws of physics are universal.

This is a statement of which law or principle?

A Hubble's law

B Doppler's principle

C Cosmological principle

D Wien's displacement law *(1 mark)*

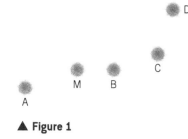

▲ Figure 1

2 Figure 1 shows five galaxies **A**, **B**, **C**, **D**, and **M**, where **M** is our galaxy (Milky Way).

Which galaxy will show the largest recession velocity when observed from **M**? *(1 mark)*

3 A star is at a distance of 2.1×10^{17} m from the Earth.

What is the distance of this star in light years (ly)?

A 4.2 ly   C 14 ly

B 6.8 ly   D 22 ly *(1 mark)*

4 The approximate wavelengths of the red-end and blue-end of the visible spectrum are 700 nm and 400 nm, respectively. The wavelength of an emission line in the blue part of the spectrum from a star shows a 0.058% increase.

What is the percentage increase in the wavelength of an emission line in the red part of the spectrum from the same star?

A 0%   C 0.058%

B 0.033%   D 0.102% *(1 mark)*

5 a With the help of a labelled diagram, show that a parallax of 1 arcsecond is equivalent to a distance of about 3.1×10^{16} m (1 pc).

1 AU = 1.5×10^{11} m *(3 marks)*

b The length of our galaxy is about 4.0×10^4 pc. Calculate the time it would take for light to travel the length of the galaxy. Write your answer both in seconds and years. *(3 marks)*

c The Hubble constant is about 70 km s^{-1} Mpc^{-1}.

i Calculate the Hubble constant in s^{-1}. *(2 marks)*

ii Estimate the farthest distance we can observe in the Universe.

Explain your answer. *(3 marks)*

d List two observations that support the Big Bang model of the Universe. *(2 marks)*

▲ Figure 2

6 Astronomers are observing the absorption lines in the visible spectrum from the star Alpha Centuari.

Figure 2 shows four data points plotted on a grid. The change in the wavelength of a spectral line is $\Delta\lambda$ and λ is wavelength of the same spectral line observed in the laboratory.

a Draw a line of best fit through the data points. *(1 mark)*

b Explain why a straight line graph is produced. *(1 mark)*

c Determine the gradient of the straight line and therefore estimate the speed of the star relative to the Earth. *(3 marks)*

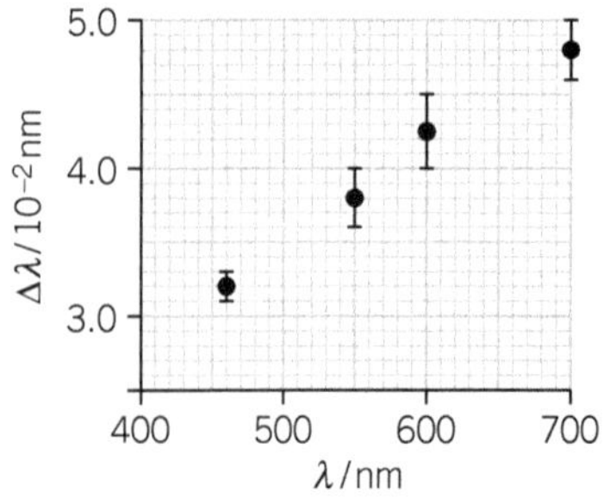

Module 6 Particles and medical physics
Chapter 21 Capacitance

In this chapter you will learn about ...

- ☐ Capacitors
- ☐ Capacitance
- ☐ Capacitors in combination
- ☐ Energy stored by capacitors
- ☐ Discharge of a capacitor
- ☐ Time constant
- ☐ Charging of a capacitor
- ☐ Uses of capacitors

21 CAPACITANCE

21.1 Capacitors
21.2 Capacitors in circuits
21.3 Energy stored by capacitors

Specification reference: 6.1.1, 6.1.2

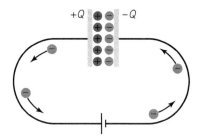

▲ Figure 1 *The plates of a capacitor get oppositely charged*

Revision tip
The farad F is a very large unit. In practice, capacitors are marked in µF, nF, and pF.
($\mu = 10^{-6}$, $n = 10^{-9}$, and $p = 10^{-12}$)

Common misconception
The letter 'C' is used for both capacitance and coulomb. There is room for confusion – you just need to be vigilant.

21.1 Capacitors

A capacitor is a component designed to store charge. It consists of two metal plates separated by an insulator (air, ceramic, paper, etc).

Figure 1 shows a capacitor connected to a source of e.m.f. (a cell). Electrons are removed from the left-hand side plate and electrons are deposited onto the opposite plate. Each plate gains and loses the same number of electrons and the final charges on the plates are $+Q$ and $-Q$. The capacitor is charged fully when the potential difference (p.d.) V across it is equal to the e.m.f. of the cell.

Capacitance

Experiments show that the charge Q on one of the capacitor plates is directly proportional to the p.d. V across the capacitor. Therefore

$$Q \propto V \quad \text{or} \quad Q = VC$$

where C is the capacitance of the capacitor. Capacitance has the unit farad F, where $1\,F = 1\,C\,V^{-1}$.

The **capacitance** of a capacitor is defined as the charge stored per unit p.d. across it.

A capacitance of 1 **farad** is defined as 1 coulomb of charge stored per unit volt.

21.2 Capacitors in circuits

In a circuit capacitors can be connected in many different combinations. You can simplify complex circuits by considering capacitors in series and parallel combinations.

Parallel

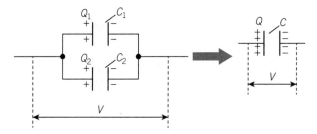

▲ Figure 2 *Parallel combination*

Figure 2 shows capacitors of capacitance C_1 and C_2 connected in parallel. This combination is equivalent to a capacitor of capacitance C.

- The p.d. V across each capacitor is the same.
- The total charge Q stored is equal to the sum of the individual charges. $Q = Q_1 + Q_2 + ...$
- The total capacitance C is given by the equation $C = C_1 + C_2 + ...$

Capacitance

Series

Figure 3 shows capacitors of capacitance C_1 and C_2 connected in series. This combination is equivalent to a capacitor of capacitance C.

- The charge Q stored by each capacitor is the same.
- The total p.d. V across the combination is the sum of the individual p.d.s. $V = V_1 + V_2 + ...$
- The total capacitance C is given by the equation

$$\frac{1}{C} = \frac{1}{C_1} + \frac{1}{C_2} + ... \quad \text{or} \quad C = (C_1^{-1} + C_2^{-1} + ...)^{-1}$$

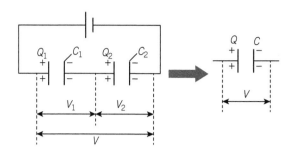

▲ **Figure 3** *Series combination*

21.3 Energy stored by capacitors

When a capacitor is charging up, work is done to push the electrons onto the negative plate and to pull the electrons away from the positive plate.

Revision tip
For capacitors in **series**, the total capacitance is always less than the *smallest* capacitance value.

Stored energy

The work done on the charges is equal to the area under a p.d. against charge graph. This area is also equal to the energy stored by the capacitor, see Figure 4.

work done W = energy stored = $\frac{1}{2}$ × charge × p.d.

$$W = \frac{1}{2}QV$$

Using $Q = VC$ gives two further equations.

$$W = \frac{1}{2}\frac{Q^2}{C} \quad \text{and} \quad W = \frac{1}{2}V^2C.$$

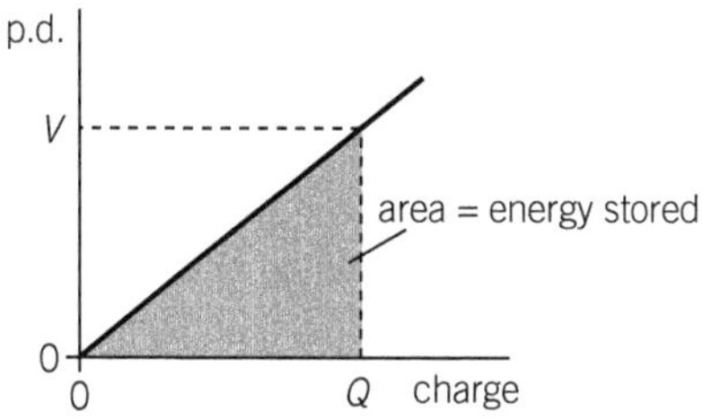

▲ **Figure 4** *The area under a p.d. against charge graph is the energy stored*

> ### 🖩 Worked example: Energy stored
>
> A 100 μF capacitor and a 220 μF capacitor are connected in **series** to a 6.0 V supply. Calculate the energy stored by the 100 μF capacitor.
>
> **Step 1:** Calculate the total capacitance.
>
> $C = (100^{-1} + 220^{-1})^{-1} = 68.75 \, \mu F$
>
> **Step 2:** Calculate the charge stored by the 100 μF capacitor.
>
> Capacitors in series store the same charge.
>
> $Q = VC = 6.0 \times 68.75 \times 10^{-6} = 4.125 \times 10^{-4} \, C$
>
> **Step 3:** Calculate the energy stored by the 100 μF capacitor.
>
> energy $= \frac{Q^2}{2C} = \frac{(4.125 \times 10^{-4})^2}{2 \times 100 \times 10^{-6}} = 8.5 \times 10^{-4} \, J$ (2 s.f.)

Summary questions

1. A 500 μF capacitor is connected to a 9.0 V supply. Calculate the charge stored by the capacitor. *(2 marks)*
2. Calculate the energy stored by the capacitor in **Q1**. *(2 marks)*
3. A student is given three 100 μF capacitors. These capacitors can be connected in any combination. Calculate the maximum and minimum capacitance. *(4 marks)*
4. Calculate the total capacitance of the circuit shown in Figure 5. *(3 marks)*

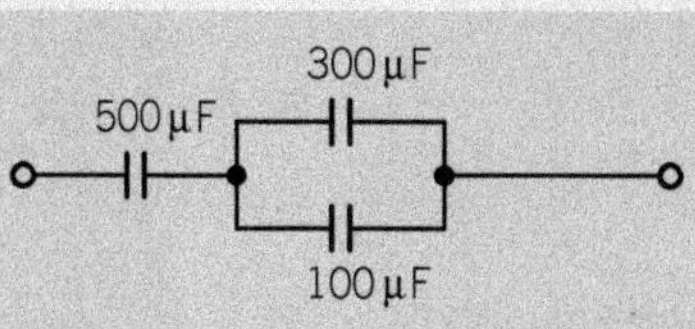

▲ **Figure 5**

5. The circuit of Figure 5 is now connected to a 6.0 V battery. Calculate the p.d. measured by a digital voltmeter placed across the 500 μF capacitor. *(3 marks)*
6. A 1000 μF capacitor has a p.d. of 10 V. It is connected across an uncharged 500 μF capacitor. Calculate the final p.d. across the 1000 μF capacitor. *(4 marks)*

21.4 Discharging capacitors
21.5 Charging capacitors
21.6 Uses of capacitors

Specification reference: 6.1.2, 6.1.3

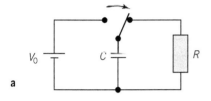

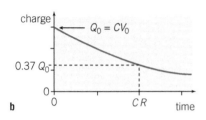

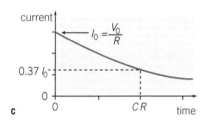

▲ **Figure 1** (a) A capacitor discharging circuit, (b) charge-time graph, and (c) current-time graph

21.4 Discharging capacitor

Knowledge of series and parallel circuits together with the three equations below can be used to analyse circuits where a capacitor discharges or charges through a resistor.

$$Q = VC \qquad V = IR \qquad I = \frac{\Delta Q}{\Delta t}$$

Exponential decay

Figure 1a shows a capacitor of capacitance C connected in **parallel** with a resistor of resistance R. The switch is connected to the left. The p.d. across the capacitor is V_0. At time $t = 0$, the switch is connected to the right. The capacitor starts to discharge through the resistor. The p.d. across the resistor and the capacitor is the same at all times.

- At $t = 0$, maximum current $I_0 = \frac{V_0}{R}$ and maximum charge $Q_0 = V_0 C$.
- The p.d, current, and charge decrease *exponentially* with respect to time.
- At time t, the p.d. V across the capacitor (or resistor), the current I in the resistor, and the charge Q on the capacitor are given by the equations

$$V = V_0 e^{-\frac{t}{CR}} \qquad I = I_0 e^{-\frac{t}{CR}} \qquad Q = Q_0 e^{-\frac{t}{CR}}$$

These equations can be represented by the equation $x = x_0 e^{-\frac{t}{CR}}$, where e is the base of natural logarithms, which has a value of 2.718... Figures 1b and 1c show the variations of charge and current with time.

Time constant

The time constant τ for a discharging capacitor is equal to the time taken for the p.d. (or the current or the charge) to decrease to e^{-1} (about 37%) of its initial value.

- The time constant τ is also equal to the product CR.
- The unit of time constant is the second (s). Note: $1\,\text{s} = 1\,\text{F}\,\Omega$.
- A capacitor never discharges fully, but after a time of $5CR$ it is 'practically discharged' with less than 1% of the original charge left.

Modelling decay

For a discharging capacitor, the p.d. across the capacitor is the same as the p.d. across the resistor.

Therefore

$$IR = -\frac{Q}{C} \qquad \text{or} \qquad R\frac{\Delta Q}{\Delta t} = -\frac{Q}{C}$$

Therefore

$$\frac{\Delta Q}{\Delta t} = -\frac{Q}{CR}$$

The minus sign signifies the charge on the capacitor decreases with time.

This equation can be used to model the discharge of a capacitor over a period of time.

Example

$CR = 2.0\,\text{s}$ and $\Delta t = 0.10\,\text{s}$

The charge lost after every $0.10\,\text{s}$ is $\Delta Q = 0.05\,Q$.

After every $0.10\,\text{s}$, 95% of the previous charge is left.

This constant-ratio property is characteristic of **exponential decay**.

21.5 Charging capacitors

Figure 2a shows a capacitor charging through a resistor. The charge Q stored by the capacitor and therefore the p.d. V_C across it *increases* with time. The p.d. V_R across the resistor (and therefore the current I) must *decrease* because $V_C + V_R = V_0$ (Kirchhoff's second law).

Important equations

- V_C and Q are given by the general equation $x = x_0(1 - e^{-\frac{t}{CR}})$
- V_R and I are given by the general equation $x = x_0 e^{-\frac{t}{CR}}$

> **Worked example: Charging capacitor**
>
> An uncharged capacitor is charged using the circuit shown in Figure 3.
>
> The switch is closed at time $t = 0$. Calculate the p.d. across the $1.0\,\mu\text{F}$ capacitor at time $t = 3.0\,\text{s}$.
>
> **Step 1:** Calculate the time constant CR of the circuit.
>
> $CR = 1.0 \times 10^{-6} \times 1.2 \times 10^{6} = 1.2\,\text{s}$.
>
> **Step 2:** Calculate the p.d. V_C across the capacitor.
>
> $V_C = V_0(1 - e^{-\frac{t}{CR}}) = 6.0 \times (1 - e^{-\frac{3.0}{1.2}}) = 5.51\,\text{V} \approx 5.5\,\text{V}$ (2 s.f.)

▲ Figure 3

21.6 Uses of capacitors

Capacitors can store energy. This energy can be released in a very short period of time to produce large power. For example, releasing 1 J of stored energy is 1 µs can produce an output power of 1 MW. Capacitors are used in camera flashes, in particle accelerators, in electrical smoothing circuits, and so on.

> **Maths: $\log_e$ or ln**
>
> ln is an abbreviation of $\log_e$
>
> $V = V_0 e^{-\frac{t}{CR}}$. Taking $\log_e$ of both sides, we get $\ln V = \ln V_0 - \frac{t}{CR}$
>
> A graph of $\ln V$ against t will be a straight line, with gradient $= -\frac{1}{CR}$.

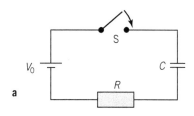

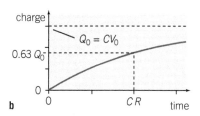

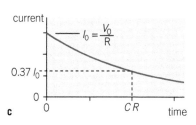

▲ **Figure 2** (a) A capacitor charging circuit, (b) charge–time graph, and (c) current–time graph

Summary questions

1. Calculate the time constant of a circuit given $C = 100\,\mu\text{F}$ and $R = 150\,\text{k}\Omega$. *(1 mark)*

2. Explain why the p.d. across a charged capacitor decreases when it is connected across a resistor. *(2 marks)*

3. In the circuit of Figure 1a, $C = 100\,\mu\text{F}$, $R = 200\,\text{k}\Omega$, and $V_0 = 10\,\text{V}$. Calculate the p.d. across the capacitor after 38 s. *(3 marks)*

4. Show that the charge left on a discharging capacitor after five time constants is less than 1%. *(3 marks)*

5. In the circuit of Figure 2a, $C = 500\,\mu\text{F}$, $R = 100\,\text{k}\Omega$, and $V_0 = 10\,\text{V}$. Calculate the p.d. across the capacitor and resistor after 80 s. *(4 marks)*

6. For the circuit in Q3, how long would it take for the p.d. across the capacitor to halve? *(4 marks)*

Chapter 21 Practice questions

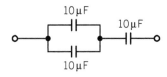

▲ Figure 1

1   Figure 1 shows a circuit with three identical capacitors.
    What is the total capacitance of the circuit?
    A   3.3 µF
    B   6.7 µF
    C   15 µF
    D   30 µF   (1 mark)

2   Which one of the following is **not** a unit for time constant?
    A   s
    B   FΩ
    C   F V A^{-1}
    D   Ω V^{-1} (1 mark)

3   A capacitor is discharging through a resistor. The time constant of the circuit is 2.0 s.
    At time $t = 0$, the p.d. across the capacitor is 8.0 V.
    What is the p.d. across the capacitor at time $t = 4.0$ s?
    A   1.1 V
    B   2.0 V
    C   2.9 V
    D   4.0 V   (1 mark)

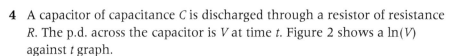

4   A capacitor of capacitance C is discharged through a resistor of resistance R. The p.d. across the capacitor is V at time t. Figure 2 shows a ln(V) against t graph.
    What is the gradient of the graph equal to?
    A -1
    B $-e$
    C $-CR$
    D $-(CR)^{-1}$ (1 mark)

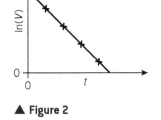

▲ Figure 2

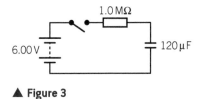

▲ Figure 3

5   Figure 3 shows a circuit with a capacitor of capacitance 120 µF and a resistor of resistance 1.0 MΩ connected via a switch to a battery of e.m.f. 6.00 V. The battery has negligible internal resistance.
    At time $t = 0$, the switch is closed and the p.d. across the capacitor is zero.
    a   Describe and explain what happens to the p.d. across the resistor and the capacitor.   (4 marks)
    b   Calculate the p.d. across the capacitor at time $t = 200$ s.   (3 marks)
    c   Calculate the maximum current in the circuit in µA.   (1 mark)
    d   Calculate the maximum energy stored by the capacitor.   (2 marks)

6   Figure 4 shows a capacitor–resistor circuit.
    The p.d. across the capacitor is 4.50 V. The switch is closed at time $t = 0$.
    a   Calculate the time constant of the circuit.   (1 mark)
    b   Calculate the p.d. across the resistor at time $t = 35$ s.   (2 marks)
    c   Calculate the total energy dissipated by the resistor between $t = 0$ and $t = 35$ s.   (3 marks)

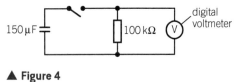

▲ Figure 4

Chapter 22 Electric fields

In this chapter you will learn about ...

- ☐ Electric field lines
- ☐ Electric field strength
- ☐ Coulomb's law
- ☐ Radial electric field
- ☐ Uniform electric field
- ☐ Capacitance of a parallel plate capacitor
- ☐ Motion of charged particles
- ☐ Electric potential
- ☐ Electric potential energy
- ☐ Capacitance of an isolated sphere

22 ELECTRIC FIELDS
22.1 Electric fields
22.2 Coulomb's law

Specification reference: 6.2.1, 6.2.2

22.1 Electric fields

A charged particle creates an electric field in the space around it. Another charged particle in this electric field will experience either an attractive or a repulsive electrical force.

Field patterns

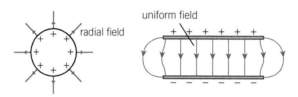

▲ **Figure 1** *Radial and uniform electric fields*

Electric field patterns can be mapped using electric field lines (or lines of force). The direction of the field at a point shows the direction of the force experience by a small *positive* charge placed at that point. The electric field strength (see later) is indicated by the separation between the field lines.

- A uniform electric field has equally spaced field lines. The electric field between two oppositely charged parallel plates is uniform.
- Electric field lines are always perpendicular to the surface of a conductor.
- A radial field has straight field lines converging to a point at the centre of the charged object. The field strength gets smaller with increased distance from the object.
- A uniformly charged sphere can be modelled as a point charge at its centre, see Figure 2.

Revision tip: Field strength
Closely spaced electric field lines indicate greater electric field strength E.

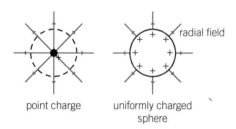

▲ **Figure 2** *A uniformly charged sphere with charge Q can be modelled as a point charge (particle) of charge Q*

Electric field strength E

The **electric field strength** E at a point is defined as the force experienced per unit *positive* charge at that point.

This can be written as

$$E = \frac{F}{Q}$$

where F is the force experienced by the positive charge Q. The SI unit for E is N C^{-1}. Electric field strength is a vector quantity.

22.2 Coulomb's law

Coulomb's law is a universal law that can be applied to all charged particles.

Coulomb's law: Two point charges exert an electrostatic (electrical) force on each other that is directly proportional to the product of their charges and inversely proportional to the square of the distance between them.

Electric fields

Equation for Coulomb's law

According to **Coulomb's law**

$$F \propto \frac{Qq}{r^2}$$

where F is the electrical force, Q and q are the charges, and r is the separation, see Figure 3. The law can be written as an equation using the permittivity of free space ε_0 (8.85×10^{-12} F m^{-1}) as follows:

$$F = \frac{Qq}{4\pi\varepsilon_0 r^2}$$

For two uniformly charged spheres, just remember that r is the centre-to-centre separation.

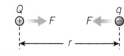

a *unlike charges attract*

b *like charges repel*

▲ **Figure 3** *Electrical forces can be either attractive or repulsive*

Radial field

The **electric field strength** E at a distance r from the centre of a charge Q can be determined as follows:

$$E = F \div q = \frac{Qq}{4\pi\varepsilon_0 r^2} \div q$$

Therefore

$$E = \frac{Q}{4\pi\varepsilon_0 r^2}$$

The electric field strength E obeys an inverse square law with distance.

Gravitational and electric fields

Here are the main similarities and differences between the electric field of a point charge and the gravitational field of a point mass.

Similarities:

- The field strengths of both obey an inverse square law with distance.

$$g \propto \frac{1}{r^2} \text{ and } E \propto \frac{1}{r^2}$$

- Both produce a radial field pattern.

Differences:

- Gravitational field depends on the mass whereas electric field depends on charge.
- A gravitational field always produces an attractive force, whereas an electric field can produce either an attractive or a repulsive force.

> **Worked example: Electron acceleration**
>
> Calculate the acceleration of an electron at a distance of 7.0×10^{-11} m from a proton.
>
> **Step 1:** Write down the quantities needed to calculate the force on the electron.
>
> $r = 7.0 \times 10^{-11}$ m $Q = q = 1.60 \times 10^{-19}$ C $\varepsilon_0 = 8.85 \times 10^{-12}$ F m^{-1}
>
> **Step 2:** Calculate the force experienced by the electron.
>
> $$F = \frac{Qq}{4\pi\varepsilon_0 r^2} = \frac{1.60 \times 10^{-19} \times 1.60 \times 10^{-19}}{4\pi \times 8.85 \times 10^{-12} \times (7.0 \times 10^{-11})^2} = 4.70 \times 10^{-8} \text{ N}$$
>
> **Step 3:** The mass of the electron is 9.11×10^{-31} kg. Use $F = ma$ to calculate the acceleration.
>
> $$a = \frac{F}{m} = \frac{4.70 \times 10^{-8}}{9.11 \times 10^{-31}} = 5.2 \times 10^{22} \text{ m s}^{-2} \text{ (2 s.f.)}$$

Summary questions

1. The electrostatic force on an electron is 8.0×10^{-14} N. Calculate the electric field strength. *(2 marks)*

2. The electric field strength at a point is 6.0×10^4 N C^{-1}. Calculate the force experienced by an electron. *(2 marks)*

3. Two electrons are separated by 2.0×10^{-10} m. Calculate the electrostatic force on one of the electrons. *(3 marks)*

4. Calculate the force on the electron in Q3 when the separation is halved. *(2 marks)*

5. A metal sphere has a positive charge of 3.8×10^{-9} C. Calculate its radius given the field strength on the surface is 5.0×10^4 N C^{-1}. *(3 marks)*

6. Charge density σ is defined as the charge per unit area. Derive an equation for σ and surface electric field strength E for a charged metal sphere. *(3 marks)*

22.3 Uniform electric fields and capacitance
22.4 Charged particles in uniform electric fields

Specification reference: 6.2.3

22.3 Uniform electric fields and capacitance

There is a uniform electric field between a pair of oppositely charged parallel plates. The plates store charge – they are equivalent to a capacitor.

Parallel plates

A charged particle of charge $+Q$ will experience a force F in the direction of the field, see Figure 1. Consider this particle moving from the positive (where it is stationary) to the negative plate. The work done on the particle is VQ, where V is the p.d. across the plates. This work done is also Fd, where d is the separation between the plates. Therefore

$$Fd = VQ$$

or

$$\frac{F}{Q} = \frac{V}{d}$$

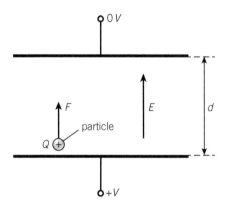

▲ **Figure 1** *A charged particle in a uniform electric field*

The electric field strength is defined as $\frac{F}{Q}$ therefore the uniform field strength of the electric field between oppositely charged plates is given by the equation

$$E = \frac{V}{d}$$

The unit for electric field strength can be either $N\,C^{-1}$ or $V\,m^{-1}$.

Capacitor with parallel plates

The capacitance C of a capacitor made from two parallel metal plates in a vacuum with separation d and area of overlap A is given by the equation

$$C = \frac{\varepsilon_0 A}{d}$$

where ε_0 is the permittivity of free space ($8.85 \times 10^{-12}\,F\,m^{-1}$). With an insulator (or dielectric) other than a vacuum between the plates, the capacitance increases and is given by the equation

$$C = \frac{\varepsilon A}{d}$$

where ε is the permittivity of the insulator. The permittivity of insulators is always greater than ε_0, so sometimes the term relative permittivity ε_r is also used, where $\varepsilon = \varepsilon_r \varepsilon_0$.

ε_r has no unit. For vacuum $\varepsilon_r = 1$ (by definition), for air $\varepsilon_r = 1.0006 \approx 1.0$, and for paper $\varepsilon_r \approx 4.0$.

Revision tip: Field strength

The defining equation for electric field strength is $E = \frac{F}{Q}$ and *not* $E = \frac{V}{d}$.

Synoptic link

Capacitance was defined in Topic 21.1, Capacitors.

22.4 Charged particles in uniform electric fields

A charged particle experiences an electrical force in a uniform electric field, and therefore will have acceleration. The motion of a charged particle in a uniform electric field can be analysed using the equations of motion and the following equations:

$$F = ma \qquad F = EQ \qquad E = \frac{V}{d}$$

Motion parallel to field

A positive charge moving in the direction of the electric field will accelerate, whereas a negative charge will decelerate when moving in the direction of the field. See Figure 2.

Motion perpendicular to field

A charged particle will describe a parabolic path when its initial velocity is perpendicular to the direction of the field. You can use your knowledge of projectiles to analyse the motion of particles. Here are a couple of important reminders:

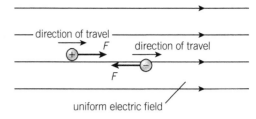

▲ **Figure 2** *Charged particles moving to the right. The positive particle will accelerate and the negative particle will decelerate*

- The velocity of the particle remains constant in a direction perpendicular to the field.
- The particle will accelerate (or decelerate) in the direction of the field.

> **Worked example: Electron deflection**
>
> An electron, with a velocity of 2.0×10^7 m s^{-1} in the horizontal direction, enters a region of uniform electric field of strength 4.0×10^4 V m^{-1}. The direction of the electric field is vertical. Calculate the vertical deflection of the electron after it has travelled 8.0 cm in the horizontal direction.
>
> **Step 1:** Calculate the time taken to travel a horizontal distance of 8.0 cm.
> $$\text{time } t = \frac{\text{distance}}{\text{speed}} = \frac{8.0 \times 10^{-2}}{2.0 \times 10^7} = 4.0 \times 10^{-9} \text{ s}$$
>
> **Step 2:** Calculate the vertical acceleration of the electron.
> $$E = 4.0 \times 10^4 \text{ V m}^{-1} \quad Q = e = 1.60 \times 10^{-19} \text{ C} \quad m = 9.11 \times 10^{-31} \text{ kg}$$
> $$a = \frac{F}{m} = \frac{EQ}{m} = \frac{4.0 \times 10^4 \times 1.60 \times 10^{-19}}{9.11 \times 10^{-31}} = 7.025 \times 10^{15} \text{ m s}^{-2}$$
>
> **Step 3:** Use the equation of motion $s = ut + \frac{1}{2}at^2$ to calculate vertical deflection s.
> $$u = 0 \quad a = 7.025 \times 10^{15} \text{ m s}^{-2} \quad t = 4.0 \times 10^{-9} \text{ s}$$
> $$s = \tfrac{1}{2} \times 7.025 \times 10^{15} \times (4.0 \times 10^{-9})^2 = 0.056 \text{ m } (5.6 \text{ cm}) \text{ (2 s.f.)}$$

Summary questions

1. Two parallel plates in air are separated by 1.0 cm. The plates are connected to a 2.0 kV supply.
   Calculate the electric field strength between the plates. *(2 marks)*
2. The capacitance of the capacitor in **Q1** is 10 pF. Calculate the charge stored by the capacitor. *(1 mark)*
3. Calculate the area of overlap of the plates in **Q1**. *(2 marks)*
4. A capacitor is made by inserting a sheet of paper of thickness 0.070 mm between A4 size metal sheets.
   Calculate the capacitance in nF of this capacitor. *(3 marks)*
   ε_r for paper = 4.0 and area of A4 = 6.24×10^{-2} m^2
5. Two metal plates are separated by 1.0 cm and are connected to a 500 V supply.
   Calculate the time it takes for an electron to travel from the negative plate to the positive plate. *(4 marks)*
   (Assume the electron is initially at rest.)
6. Two parallel plates are charged oppositely using a power supply. The power supply is then disconnected. The energy stored by the capacitor is E_0. The separation between the charged plates is doubled.
   Calculate the final energy stored in terms of E_0. *(3 marks)*

22.5 Electric potential and energy

Specification reference: 6.2.4

22.5 Electric potential and energy

You need the idea of **electric potential** to understand the gain and loss in the electric potential energy of charged particles.

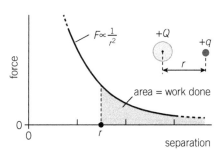

▲ **Figure 1** *The area under a force–separation graph is equal to the work done on the charges*

Potential energy

Consider two charged particles of charges Q and q with a separation r. The force experienced by each particle is given by the equation for Coulomb's law

$$F = \frac{Qq}{4\pi\varepsilon_0 r^2}$$

where ε_0 is the permittivity of free space ($8.85 \times 10^{-12}\,\text{F m}^{-1}$).

External work must be done to bring like charged particles closer because they repel each other. The work done can be calculated from a force against separation graph, see Figure 1.

The area under a force–separation graph is the work done on the charged particles. The total work done on the charged particles to bring them from infinity to a separation r is known as electric potential energy.

The electric potential energy E for two charged particles separated by a distance r is defined as the work done to bring the charged particles from infinity to a separation r.

The equation for electric potential energy is $E = \frac{Qq}{4\pi\varepsilon_0 r^2}$.

- This equation can be used for uniformly charged spheres – just remember that r will be the separation between the centres of the spheres.
- The electric potential energy is negative for two particles with unlike charges. The negative potential energy simply means that external work has to be done to pull apart the particles.
- The *change* in electric potential energy ΔE is simply calculated as follows:

$$\Delta E = \text{final } E - \text{initial } E = \frac{Qq}{4\pi\varepsilon_0}\left(\frac{1}{r_f} - \frac{1}{r_i}\right)$$ where r_f is the final separation and r_i is the initial separation.

Electric potential

The **electric potential** V at a point is defined as the work done per unit charge in bringing a positive charge from infinity to that point.

The electric potential V at a distance r away from a particle of charge Q can be determined by dividing the electric potential energy by the test charge q. This gives the following equation for electric potential:

$$V = \frac{Q}{4\pi\varepsilon_0 r}$$

- V has unit J C^{-1} or volt (V).
- V is defined to be zero at infinity.
- This equation can be used for a uniformly charged sphere – just remember that r will be the distance from the centre of the sphere.
- The *change* in electric potential ΔV is simply calculated as follows:

$$\Delta V = \text{final } V - \text{initial } V = \frac{Q}{4\pi\varepsilon_0}\left(\frac{1}{r_f} - \frac{1}{r_i}\right)$$

where r_f is the final distance and r_i is the initial distance.

> **Revision tip**
> Electric potential energy = electric potential × charge
> Therefore, $E = Vq$.

Electric fields

Capacitance revisited

An isolated metal sphere can be charged using a power supply. The electric potential V on the surface of the sphere is related to the surface charge Q and the radius R of the sphere by the equation

$$V = \frac{Q}{4\pi\varepsilon_0 R}$$

The ratio $\frac{Q}{V}$ is the capacitance C of the sphere. Therefore

$$C = 4\pi\varepsilon_0 R$$

The value of $4\pi\varepsilon_0$ is $1.11 \times 10^{-10}\,\text{F m}^{-1}$. Therefore a sphere of radius 1.0 m will have capacitance of about 11 pF.

Worked example: Removing an electron

In the hydrogen atom, the electron is at a mean distance of about $5.0 \times 10^{-11}\,\text{m}$ from the proton.

Estimate the energy required to completely remove the electron from the hydrogen atom in eV.

Step 1: Write down the information given.

$Q = +1.60 \times 10^{-19}\,\text{C}$ $q = -1.60 \times 10^{-19}\,\text{C}$ $r = 5.0 \times 10^{-11}\,\text{m}$ $\varepsilon_0 = 8.85 \times 10^{-12}\,\text{F m}^{-1}$

Step 2: Calculate the electric potential energy of the electron and proton.

$$E = \frac{Qq}{4\pi\varepsilon_0 r} = -\frac{(1.60 \times 10^{-19})^2}{4\pi \times 8.85 \times 10^{-12} \times 5.0 \times 10^{-11}} = -4.60 \times 10^{-18}\,\text{J}$$

Step 3: Calculate the energy required to completely remove the electron in eV.

$1\,\text{eV} = 1.60 \times 10^{-19}\,\text{J}$

$$\text{energy} = \frac{4.60 \times 10^{-18}}{1.60 \times 10^{-19}} = 29\,\text{eV (2 s.f.)}$$

Revision tip

The electron and the proton have a charge of the same magnitude, $1.60 \times 10^{-19}\,\text{C}$.

Summary questions

1. The electric potential on the surface of a sphere is −100 V.
   What is the energy required to remove a charge of 1 C from the surface of the sphere to infinity? *(1 mark)*
2. The electric potential on the surface of a sphere is +2000 V.
   Calculate the potential at a distance equal to one radius from the *surface* of the sphere. *(2 marks)*
3. Calculate the electric potential at a distance of $1.2 \times 10^{-10}\,\text{m}$ from a proton. *(2 marks)*
4. The electric potential on the surface of a metal sphere of radius 5.0 cm is 100 V.
   Calculate the charge on the sphere. *(3 marks)*
5. Calculate the capacitance of a metal sphere of radius 2.5 cm and the charge stored on its surface when it is connected to the positive terminal of a 5.0 kV supply. *(4 marks)*
6. Estimate the force experienced by each charged metal sphere shown in Figure 2. The electric potential on the surface of each sphere is +3000 V. *(4 marks)*

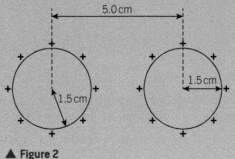

▲ Figure 2

Chapter 22 Practice questions

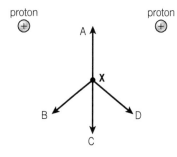
▲ Figure 1

1 Figure 1 shows two protons. Point **X** is at the same distance from each particle. Which arrow shows the correct direction of the resultant electric field at **X**? *(1 mark)*

2 What is electric potential measured in base units?
   A $kg\,m^2\,A^{-1}s^{-3}$
   B $kg\,m^{-2}\,A^{-1}s^{-3}$
   C $kg\,m^2\,s^3\,A^{-1}$
   D $kg\,m^2\,A\,s^{-3}$ *(1 mark)*

3 An electron experiences a force of magnitude F in a uniform electric field. What is the force experienced by an alpha particle, in terms of F, in the same electric field?
   A 0
   B F
   C $2F$
   D $4F$ *(1 mark)*

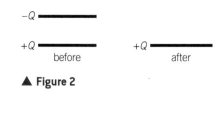
▲ Figure 2

4 Figure 2 shows opposite charged plates before and after the separation between the plates is doubled. The initial energy stored by this arrangement is E.
   What is the final energy stored in terms of E?
   A $\dfrac{E}{2}$
   B E
   C $2E$
   D $4E$ *(1 mark)*

5 A metal sphere is momentarily connected to the positive terminal of a high-voltage power supply. The radius of the sphere is 5.0 cm. A scientist determines the electric field strength E at a distance r from the centre. The results are recorded in the table below.

r / cm	9.2	18.0	25.0
E / $10^3\,N\,C^{-1}$	20.7	5.4	2.8

   a Use the table to confirm the relationship between E and r. *(2 marks)*
   b Calculate the charge Q on the surface of the sphere. *(3 marks)*
   c Calculate the charge stored per unit surface area. *(2 marks)*
   d Calculate the electric potential on the surface of the sphere. *(2 marks)*

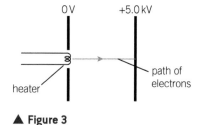
▲ Figure 3

6 Figure 3 shows two short parallel plates used to accelerate electrons from a heater.
   The potential difference between the plates is 5.0 kV and the separation between the plates is 7.2 cm.
   a Describe the electric field and the field strength between the plates. *(2 marks)*
   b Calculate the electric field strength between the plates. *(2 marks)*
   c Calculate the maximum speed of an electron when at the positive plate. Assume the initial speed of the electron is zero. *(4 marks)*
   d The p.d. across the plates is fixed at 5.0 kV.
      Explain how the final kinetic energy of an electron arriving at the positive plate depends on the separation between the plates. *(2 marks)*

Chapter 23 Magnetic fields

In this chapter you will learn about ...

- ☐ Magnetic field lines
- ☐ Electromagnetism
- ☐ Magnetic flux density
- ☐ Fleming's left-hand rule
- ☐ Motion of charged particles
- ☐ Velocity selector
- ☐ Electromagnetic induction
- ☐ Faraday's law
- ☐ Lenz's law
- ☐ A.C. generator
- ☐ Transformers

23 MAGNETIC FIELDS
23.1 Magnetic fields
23.2 Understanding magnetic fields

Specification reference: 6.3.1

23.1 Magnetic fields

Magnetic fields are created in the space around a permanent magnet and a current-carrying conductor.

Field patterns

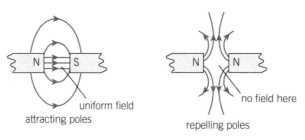

▲ Figure 1 *Field patterns between unlike poles and like poles*

> **Revision tip: Field strength**
> A uniform magnetic field has equally spaced and parallel field lines. Closely spaced magnetic field lines indicate greater magnetic flux density B.

Magnetic field patterns can be mapped using **magnetic field lines**, see Figure 1. The direction of the field at a point shows the direction of the force experienced by a free north pole at that point.

Electromagnetism

A magnetic field is produced by moving charged particles, for example electrons moving within a wire.

Figure 2 shows the magnetic field patterns for a long straight current-carrying conductor, a flat coil, and a long solenoid.

- The magnetic fields lines are concentric circles for a long straight current-carrying conductor.
- The magnetic field pattern of a solenoid resembles the field pattern of a bar magnet. The magnetic field is uniform within the core of the solenoid.

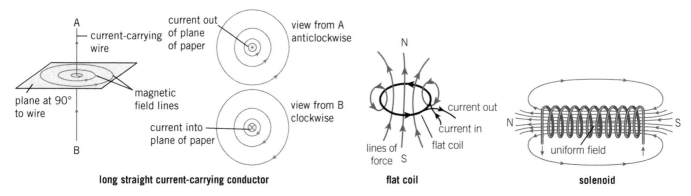

▲ Figure 2 *Magnetic field patterns for a long straight current-carrying conductor, a flat coil, and a long solenoid*

23.2 Understanding magnetic fields

The strength of a magnetic field is called magnetic flux density. This important quantity is defined in terms of the force experienced by a current-carrying conductor.

Magnetic fields

Force on a current-carrying conductor

A current-carrying wire experiences a force when it is placed at right angles to the magnetic field of a magnet. According to Newton's third law the wire and the magnet experience equal but opposite forces.

The direction of the force experienced by the current-carrying wire in a uniform field can be determined using **Fleming's left-hand rule**, see Figure 3.

Magnetic flux density and $F = BIL$

The force F experienced by the wire placed in a uniform magnetic field is given by the relationship

$$F \propto IL \sin\theta$$

where L is the length of the wire in the field, I is the current in the wire, and θ is the angle between the field and the wire. The constant of proportionality is called the magnetic flux density B of the magnetic field. Therefore

$$F = BIL \sin\theta$$

The SI unit for magnetic flux density is the **tesla** (T) and $1\,T = 1\,N\,m^{-1}\,A^{-1}$. Magnetic flux density is a vector quantity.

The **magnetic flux density** is 1 tesla when a wire carrying a current of 1 ampere placed perpendicular to the magnetic field experiences a force of 1 newton per metre of its length.

When the wire is perpendicular to the magnetic field, $\theta = 90°$ and $\sin\theta = 1$. Therefore

$$F = BIL$$

You can determine the magnetic flux density between the poles of a magnet using the arrangement shown in Figure 4.

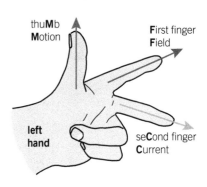

▲ **Figure 3** *Fleming's left-hand rule: The first finger points in the direction of the external field, the second finger points in the direction of the **conventional** current, and the thumb points in the direction of the motion (force)*

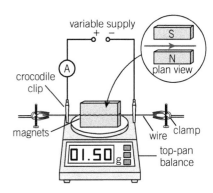

▲ **Figure 4** *An arrangement for determining magnetic flux density. The magnet experiences an equal but opposite force to the current-carrying wire placed between the poles*

 Worked example: Magnetic flux density

A copper wire of length 6.0 cm and carrying a current of 8.0 A is placed in a uniform magnetic field. The maximum magnetic force experienced by the wire is 2.9×10^{-2} N. Calculate the magnetic flux density.

Step 1: Write the quantities given in the question.

$F = 2.9 \times 10^{-2}\,N \qquad B = ? \qquad I = 8.0\,A \qquad L = 0.060\,m$

Step 2: Use the equation $F = BIL$ to calculate B.

$$B = \frac{F}{IL} = \frac{2.9 \times 10^{-2}}{8.0 \times 0.060} = 6.0 \times 10^{-2}\,T\,(2\,s.f.)$$

Summary questions

1. What can you infer about the magnetic flux density in a region that has equally spaced and parallel magnetic field lines? *(1 mark)*
2. Describe the magnetic field of a solenoid. *(2 marks)*

3. A wire of length 4.0 cm is placed perpendicular to a uniform magnetic field of flux density 0.020 T. The current in the wire is 5.0 A. Calculate the magnetic force experienced by the wire. *(2 marks)*
4. A current-carrying wire experiences a force F in a uniform magnetic field. The magnetic flux is trebled and the current in the wire is halved. Determine the new force on the wire in terms F. Explain your answer. *(2 marks)*

5. Two current-carrying copper wires are placed parallel to each other. Explain why each will experience a magnetic force. *(1 mark)*
6. The wire in Q3 is now placed at an angle θ to the magnetic field. The magnetic force on the wire is 1.5 mN. Calculate θ. *(2 marks)*

23.3 Charged particles in magnetic fields

Specification reference: 6.3.2

23.3 Charged particles in magnetic fields

A current-carrying wire in a uniform magnetic field experiences a force because each electron moving in the wire experiences a tiny force.

Magnetic fields are used in many devices, from oscilloscopes to particle accelerators, to control the paths of charged particles.

Circular paths

Figure 1 shows the path of a single electron as it travels through a region of uniform magnetic field. The magnetic field is into the plane of the paper. The electron describes a circular path within the magnetic field because the magnetic force F it experiences is perpendicular to its velocity.

No work is done by the magnetic field on the electron because the force is perpendicular to the velocity – this force has no component in the direction of travel and therefore the speed of the electron remains constant.

The magnitude of the magnetic force F experienced by a particle of charge Q moving at right angles to the magnetic field is given by the equation

$$F = BQv$$

where B is the magnetic flux density and v is the speed of the particle.

For an electron, the charge Q is the elementary charge $e = 1.60 \times 10^{-19}$ C. Therefore, the equation for the force F for an electron becomes

$$F = Bev$$

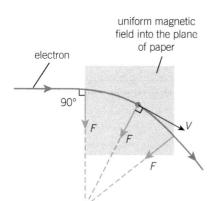

▲ **Figure 1** *The force F on the electron is perpendicular to its velocity v*

Revision tip

When using Fleming's left-hand rule, do remember that the second finger shows the direction of *conventional* current and that this direction is opposite to the direction of electron flow.

Synoptic link

Centripetal force and circular motion were covered in Topic 16.3, Exploring centripetal forces.

Going round

Consider a charged particle of mass m and charge Q moving perpendicular to a uniform magnetic field of flux density B. The particle will describe a circular path of radius r. The magnetic force BQv provides the centripetal force, therefore

$$BQv = \frac{mv^2}{r} \quad \text{or} \quad r = \frac{mv}{BQ}$$

You can also use the equation $v = \frac{2\pi r}{T}$, where T is the period, to analyse the motion of the particle.

 Worked example: Circular tracks

Electrons are travelling in a circular path at right angles to a uniform magnetic field.

Show that the period of revolution T of an electron is independent of the radius of the path or its speed, but depends on its mass m, the magnetic flux density B, and the elementary charge e.

Step 1: Write an expression for the centripetal force on the electron.

The centripetal force is provided by the BQv force. Therefore

$$BQv = \frac{mv^2}{r} \quad \text{or} \quad BQ = \frac{mv}{r}$$

Step 2: Substitute $v = \frac{2\pi r}{T}$ and $Q = e$ into the expression above.

$$Be = \frac{m}{r} \times \frac{2\pi r}{T} = \frac{2\pi m}{T}$$

Step 3: Simplify the expression and derive an equation for T.

$$T = \frac{2\pi m}{Be}$$

The period T just depends on m, B, and e.

Magnetic fields

Velocity selector

A velocity selector is a device used in instruments such as a mass spectrometer to select charged particles of a specific speed. Figure 2 shows a simple velocity selector.

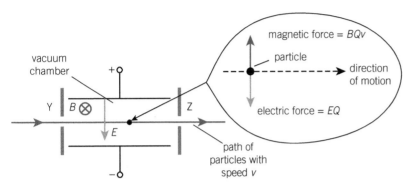

▲ **Figure 2** *Velocity selector has crossed electric and magnetic fields*

Two oppositely charged parallel plates provide a uniform electric field of field strength E. There is also a uniform magnetic field of flux density B at right angles to the electric field.

- The electric force on a charged particle of charge Q is equal to EQ.
- The magnetic force on the charged particle is equal to BQv, where v is the speed of the particle.
- The magnitude of either E or B is adjusted so that the magnetic force and the electric force are equal in magnitude and in opposite directions. Therefore

$$EQ = BQv \quad \text{or} \quad v = \frac{E}{B}$$

Any charged particle with the right speed v will travel in a straight line and emerge from the slit Z.

Summary questions

1. Explain why the speed of the electron in Figure 1 does not change. *(2 marks)*

2. Calculate the maximum magnetic force on an electron travelling at a speed of $4.0 \times 10^6 \, \text{m s}^{-1}$ in a uniform magnetic field of flux density $0.080\,\text{T}$. *(2 marks)*

3. Calculate the acceleration of the electron in **Q2**. *(2 marks)*
4. Use your answer to **Q3** to calculate the radius of the circular path described by the electron. *(2 marks)*

5. An ion of mass $6.7 \times 10^{-27}\,\text{kg}$, charge $+2e$, and speed $6.0 \times 10^5 \,\text{m s}^{-1}$ describes a circular path in a uniform magnetic field of flux density $720\,\text{mT}$. Calculate the radius of the path. *(3 marks)*
6. Calculate the period of revolution of a proton travelling at right angles to a magnetic field of magnetic flux density $900\,\text{mT}$. *(3 marks)*

23.4 Electromagnetic induction
23.5 Faraday's law and Lenz's law
23.6 Transformers

Specification reference: 6.3.3

23.4 Electromagnetic induction

An electromotive force (e.m.f.) is induced in a coil whenever there is relative motion between a magnet and the coil. This is an example of electromagnetic induction.

Important definitions

In order to understand electromagnetic induction, you need to know about magnetic flux density B (see Topic 23.1, Magnetic fields, and Topic 23.2, Understanding magnetic fields), magnetic flux ϕ, and magnetic flux linkage.

The **magnetic flux** ϕ is defined as the product of the component of the magnetic flux density B perpendicular to the area and the cross-sectional area A.

$$\phi = BA\cos\theta$$

where θ is the angle between the normal to the coil and the magnetic field, see Figure 1. When the magnetic field is normal to the area, $\cos 0° = 1$ and $\phi = BA$. The SI unit for magnetic flux is the weber (Wb) and $1\,\text{Wb} = 1\,\text{T}\,\text{m}^2$.

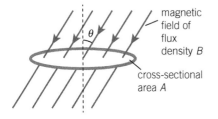

▲ **Figure 1** *Magnetic flux $\phi = BA\cos\theta$*

Magnetic flux linkage is defined as the product of the number of turns in the coil N and the magnetic flux.

magnetic flux linkage = $N\phi$

The SI unit of magnetic flux linkage is also the weber, but sometimes weber-turns (Wb-turns) is used to avoid confusion with magnetic flux.

23.5 Faraday's law and Lenz's law

An e.m.f. is induced in a circuit whenever there is a *change* in the magnetic flux linkage.

The laws

Faraday's law: The magnitude of the induced e.m.f. is directly proportional to the rate of change of magnetic flux linkage.

$$\varepsilon \propto \frac{\Delta(N\phi)}{\Delta t}$$

where ε is the induced e.m.f and $\Delta(N\phi)$ is the change in magnetic flux linkage in a time interval Δt.

This relationship can be written as an equation with the constant of proportionality equal to -1. Therefore

$$\varepsilon = -\frac{\Delta(N\phi)}{\Delta t}$$

The minus sign is a mathematical way of showing that energy is conserved as stated by Lenz's law.

Lenz's law: The direction of the induced e.m.f. or current is always such as to oppose the change producing it.

A.C. generator

Figure 2 shows a simple alternating current generator. A coil is rotated at a steady speed in a magnetic field. The magnetic flux linking the coil varies

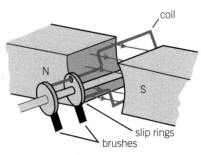

▲ **Figure 2** *A simple A.C. generator*

sinusoidally with time. The induced e.m.f. ε in the coil can be determined using Faraday's law, see Figure 3.

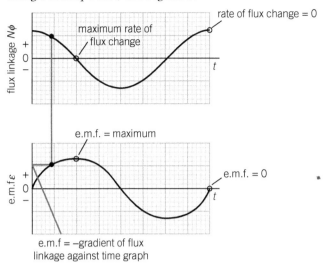

▲ Figure 3 *The variation of flux linkage with time (above) and of the induced e.m.f. with time (below)*

Worked example: Induced e.m.f in a search coil

A search coil has 2000 turns and a cross-sectional area of $6.4 \times 10^{-3}\,m^2$. It is placed perpendicular to a magnetic field. The magnetic flux density changes from 60 mT to 20 mT in a time of 0.30 s. Calculate the magnitude of the induced e.m.f. in the coil.

Step 1: Write down the quantities given.

initial $B = 60\,mT$ final $B = 20\,mT$ $A = 6.4 \times 10^{-3}\,m^2$ $N = 2000$ turns $\Delta t = 0.30\,s$

Step 2: Use Faraday's law to determine the induced e.m.f. ε.

$$\varepsilon = -\frac{\Delta(N\phi)}{\Delta t} = \frac{[2000 \times 20 \times 10^{-3} \times 6.4 \times 10^{-3}] - [2000 \times 60 \times 10^{-3} \times 6.4 \times 10^{-3}]}{0.30 - 0}$$

$\varepsilon = 1.7\,V$ (2 s.f.)

23.6 Transformers

Figure 4 shows a simple transformer. An alternating current is supplied to the primary (input) coil. This produces a varying magnetic flux in the soft iron core. The iron core ensures that all the magnetic flux created by the primary coil links the secondary (output) coil and none is lost. A varying e.m.f. is produced across the ends of the secondary in accordance with Faraday's law.

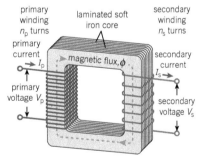

▲ Figure 4 *A transformer has two coils mounted onto a laminated soft iron core*

Transformers are used to step-up and step-down voltages.

Important equations

The input voltage V_p and the output voltage V_s are related to the number of turns on the primary coil n_p and number of turns on the secondary coil n_s by the **turn-ratio equation**

$$\frac{n_s}{n_p} = \frac{V_s}{V_p}$$

For a 100% efficient transformer

output power = input power

$$V_s I_s = V_p I_p \quad \text{or} \quad \frac{V_s}{V_p} = \frac{I_p}{I_s}$$

where I_p is the current in the primary coil and I_s is the current in the secondary coil.

Summary questions

1. What is 1 Wb in terms of T and m? *(1 mark)*

2. The output voltage from a step-up transformer is greater than the input voltage. State which of the coils, primary or secondary, has the greater number of turns. *(1 mark)*

3. A bar magnet is placed inside the core of a coil and is at rest. Explain why there is no induced e.m.f. in the coil. *(2 marks)*

4. A transformer changes a voltage of 230 V to 5.0 V. Compare the number of turns used to make the primary and secondary coils. *(2 marks)*

5. Explain how increasing the frequency of rotation would affect the induced e.m.f. in a generator. *(3 marks)*

6. A coil with 500 turns and a cross-sectional area of $2.0 \times 10^{-4}\,m^2$ is placed perpendicular to a magnetic field. The magnetic flux density is reduced to zero in a time of 0.20 s. The average induced e.m.f. in the coil is 10 mV. Calculate the initial magnetic flux density through the coil. *(3 marks)*

Chapter 23 Practice questions

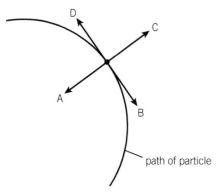

▲ Figure 1

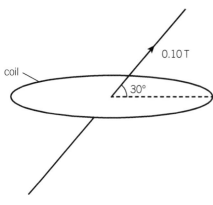

▲ Figure 2

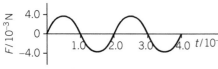

▲ Figure 3

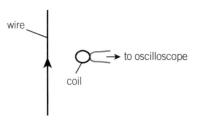

▲ Figure 4

1. A charged particle describes a circular path in a uniform magnetic field. Figure 1 shows part of this path.
   Which arrow shows the correct direction of the magnetic force acting on the particle? *(1 mark)*

2. A lithium nucleus $^{7}_{3}$Li travelling at 2.0×10^{5} m s^{-1} enters a region of uniform magnetic field of flux density 0.10 T.
   What is the maximum magnetic force experienced by this nucleus?
   A 3.2×10^{-15} N
   B 9.6×10^{-15} N
   C 2.2×10^{-14} N
   D 3.2×10^{-14} N *(1 mark)*

3. Figure 2 shows a coil with 200 turns and cross-sectional area 2.0×10^{-3} m^{2} placed in a uniform magnetic field of flux density 0.10 T. The field makes an angle of 30° to the plane of the coil.
   What is the magnetic flux linkage for this coil?
   A 2.0×10^{-4} Wb
   B 2.0×10^{-2} Wb
   C 3.5×10^{-2} Wb
   D 4.0×10^{-2} Wb *(1 mark)*

4. A copper wire is placed in a uniform magnetic field. The magnetic field is at right angles to the wire. The magnetic flux density is 60 mT. The current in the wire is alternating at a frequency f. Figure 3 shows the variation of the force F experienced by the 1.0 cm length of the wire with time t.
   a Determine the frequency f of the current in the wire. *(1 mark)*
   b Calculate the maximum current in the wire. *(2 marks)*
   c Explain how the graph in Figure 3 would change when the angle between the wire and the magnetic field is slowly reduced. *(2 marks)*

5. A charged particle enters a region of uniform magnetic field. The magnetic field is perpendicular to the velocity of the particle. The particle describes a circular arc in the magnetic field of radius r.
   a Show that the momentum p of the charged particle is given by the equation
   $$p = BQr$$
   where B is the magnetic flux density and Q is the charge of the particle. *(2 marks)*
   b An electron describes a circle of radius 3.2 cm in a magnetic field of flux density 0.012 T.
   Calculate:
   i the momentum of the electron; *(2 marks)*
   ii the kinetic energy of the electron in eV. *(3 marks)*

6. Figure 4 shows a search coil placed close to a wire. The coil is connected to an oscilloscope.
   a Explain why an e.m.f. lasting a short period of time is produced when the current in the wire is switched on. *(3 marks)*
   b With the help of a diagram, sketch the magnetic field surrounding the current-carrying wire. *(2 marks)*

Chapter 24 Particle physics

In this chapter you will learn about ...

- ☐ Scattering of alpha particles
- ☐ The nucleus
- ☐ Model of the atom
- ☐ Radius of the nucleus
- ☐ Density of the atom and the nucleus
- ☐ Strong nuclear force
- ☐ Weak nuclear force
- ☐ Particles and antiparticles
- ☐ Hadrons and leptons
- ☐ Quarks
- ☐ Beta decay

24 PARTICLE PHYSICS
24.1 Alpha-particle scattering experiment
24.2 The nucleus

Specification reference: 6.4.1

24.1 Alpha-particle scattering experiments

Before 1911 physicists knew about atoms and also knew their diameter to be about 10^{-10} m. Then, experiments carried out by Rutherford, Geiger, and Marsden on the scattering of alpha particles by thin metal foils provided evidence for the nuclear model of the atom.

Rutherford's scattering experiment

Figure 1 shows the arrangement used to perform the scattering experiments. A beam of alpha particles from a radioactive source were targeted towards a thin gold foil.

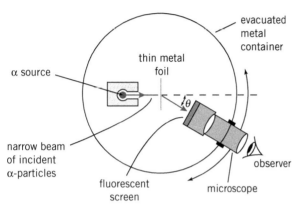

▲ **Figure 1** *Apparatus used to scatter alpha particles*

▼ **Table 1**

Observations	Conclusions
Most of the alpha particles went straight through the gold foil.	The gold atoms were mostly empty space (vacuum).
Some of the alpha particles were scattered through large angles (> 90°).	Each gold atom has a positive nucleus with a radius less than about 10^{-14} m*. (*A better modern approximation for the radius of the nucleus is about 10^{-15} m.)

An alpha particle is a helium nucleus with a charge of $+2e$, where e is the elementary charge. The charge on the gold nucleus is $+79e$. The scattering of an alpha particle can be explained in terms of electrostatic repulsion by the gold nucleus. The force on the particle can be calculated using the equation $F = \dfrac{Qq}{4\pi\varepsilon_0 r^2}$. Figure 2 shows the paths of some alpha particles.

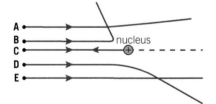

▲ **Figure 2** *Paths of alpha particles. The alpha particle at C is back-scattered and the alpha particle at E is too far away and therefore is not deflected*

24.2 The nucleus

The nucleus has neutrons and protons. The neutron has no charge and the proton has a positive charge $+e$. The term **nucleon** is used to refer to either a proton or a neutron. The nucleus is surrounded by a cloud of electrons. A neutral atom has an equal number of protons and electrons.

> **Revision tip**
> The diameter of the atom is about 10^{-10} m and the nucleus is 100 000 times smaller with a diameter of about 10^{-15} m.

Model of the atom

The nucleus of an atom for a particular element is represented as $^A_Z X$, where X is the chemical symbol for the element (e.g., Pb for lead), A is the **nucleon number** (the total number of protons and neutrons), and Z is the **proton number** or the **atomic number**. The number of neutrons is equal to $(A - Z)$.

Isotopes are nuclei of the same element that have the same number of protons but different numbers of neutrons.

Some of the isotopes of oxygen: $^{14}_8O$ $^{15}_8O$ $^{16}_8O$ $^{17}_8O$ $^{18}_8O$ $^{19}_8O$

- The radius R of a nucleus is given by the equation $R = r_0 A^{\frac{1}{3}}$ where A is the nucleon number and r_0 is a constant with a value of about 1.2 fm or 1.2×10^{-15} m.

Particle physics

- In nuclear physics, the mass of a particle is often quoted in **atomic mass units** u instead of kg.

$$1\,u = 1.661 \times 10^{-27}\,kg.$$

The masses of the electron, proton, and neutron are 0.00055 u, 1.00728 u, and 1.00867 u, respectively.

Density

The volume of the nucleus is about 10^{15} times smaller than the volume of the atom. The mass of the **atom** is approximately the same as the mass of the **nucleus** because of the miniscule mass of the electrons. The density of a material is $\frac{\text{mass}}{\text{volume}}$. Therefore, the mean density of the nucleus must be about 10^{15} times the mean density of atoms. The density of nuclear matter is about $10^{17}\,kg\,m^{-3}$.

Strong nuclear force

The protons within a nucleus experience large electrostatic repulsive forces. The attractive gravitational force between the protons is too small to explain why they remain in the nucleus. All nucleons inside the nucleus experience the **strong nuclear force**. The strong force is attractive to about 3 fm and repulsive below about 0.5 fm, see Figure 3.

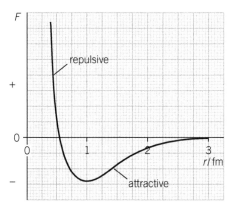

▲ **Figure 3** *Graph showing the variation of the nuclear force between two nucleons with separation r*

Worked example: Very dense

Estimate the density of a proton.

Step 1: Estimate the radius of the proton.

$R = r_0 A^{\frac{1}{3}}$; $A = 1$ for the proton therefore radius $= r_0 = 1.2 \times 10^{-15}\,m$

Step 2: Estimate the density.

mass of proton $\approx 1.7 \times 10^{-27}\,kg \qquad$ volume $= \frac{4}{3}\pi R^3$

density $= \frac{\text{mass}}{\text{volume}} = \frac{1.7 \times 10^{-27}}{\frac{4}{3}\pi \times (1.2 \times 10^{-15})^3} \approx 2 \times 10^{17}\,kg\,m^{-3}$ (1 s.f.)

Summary questions

1. State how many protons and neutrons there are in an oxygen-15 ($^{15}_{8}O$) nucleus. *(2 marks)*

2. Explain why very few alpha particles are scattered through large angles in the scattering experiment. *(2 marks)*

3. Calculate the radius in fm of the $^{4}_{2}He$ nucleus and the $^{235}_{92}U$ nucleus. *(2 marks)*

4. Estimate the number of atoms in a chain of atoms of length 1.0 cm. *(2 marks)*

5. An alpha particle with initial kinetic energy of 7.7 MeV makes a head-on collision with a nucleus of aluminium. The proton number for the aluminium nucleus is 13. Calculate the minimum separation between these two particles. *(4 marks)*

6. In Q5, calculate the electrostatic force experienced by the alpha particle at the minimum separation. *(3 marks)*

24.3 Antiparticles, hadrons, and leptons
24.4 Quarks
24.5 Beta decay

Specification reference: 6.4.2

24.3 Antiparticles, hadrons, and leptons

Most of the matter in the Universe is made up of matter particles. Antimatter particles also exist, but in smaller numbers. An electron is a matter particle and the **positron** is its corresponding antimatter. The positron has the same mass as an electron but, unlike the electron, it has a positive charge of $+e$. When a particle and its **antiparticle** meet, they completely destroy each other and produce high-energy photons. Such an event is called **annihilation**.

The list below shows some of the particle-antiparticle pairs you will come across in this course:

electron–positron   proton–antiproton   neutron–antineutron
neutrino–antineutrino

Fundamental particles

A **fundamental particle** has no internal structure and therefore cannot be divided into smaller particles. The electron is a fundamental particle. Quarks *are* considered to be fundamental particles but protons and neutrons are *not* fundamental particles because they are composed of quarks.

Particles are classified into two families – **hadrons** and **leptons**.

- Hadrons are particles and antiparticles that are affected by the **strong nuclear force**. Examples include protons, neutrons, baryons, and mesons. Charged hadrons also experience the electromagnetic force. Hadrons decay by the **weak nuclear force**.
- Leptons are particles and antiparticles that are not affected by the strong nuclear force. Examples include electrons, neutrinos, and muons. Charged leptons will also experience the electromagnetic force.

Table 1 shows the characteristics of the four fundamental forces or interactions.

▼ **Table 1** *The four fundamental forces or interactions*

Fundamental force	Effect	Relative strength	Range
strong nuclear	experienced by nucleons	1	~ 10^{-15} m
electromagnetic	experienced by static and moving charged particles	10^{-3}	infinite
weak nuclear	responsible for beta-decay	10^{-6}	~ 10^{-18} m
gravitational	experienced by all particles with mass	10^{-40}	infinite

24.4 Quarks

Any particle made up of **quarks** is called a hadron. The standard model of elementary particles requires six quarks (up, down, charm, strange, top, and bottom or simply u, d, c, s, t, and b) and their six anti-quarks to explain the existence of all the hadrons. All quarks have a charge Q that is a fraction of the elementary charge e. For example, the down quark has a charge $-\frac{1}{3}e$ or for simplicity just $-\frac{1}{3}$.

> **Synoptic link**
>
> There is more information on annihilation in Topic 26.1, Einstein's mass–energy equation.

Particle physics

▼ **Table 2** *A summary of the three quarks you need for this course*

Quark			Anti-quark		
Name	Symbol	$\dfrac{\text{Charge } Q}{e}$	Name	Symbol	$\dfrac{\text{Charge } Q}{e}$
up	u	$+\tfrac{2}{3}$	anti-up	$\bar{u}$	$-\tfrac{2}{3}$
down	d	$-\tfrac{1}{3}$	anti-down	$\bar{d}$	$+\tfrac{1}{3}$
strange	s	$-\tfrac{1}{3}$	anti-strange	$\bar{s}$	$+\tfrac{1}{3}$

> **Revision tip**
> A bar over the letter for the particle is used to denote most of the antiparticles.

Protons and neutrons

As mentioned earlier, protons and neutrons are hadrons. The hadron group is further divided into **baryons** and **mesons**, see Figure 1. Baryons have a combination of three quarks and mesons have a combination of a quark and an anti-quark.

- A proton has two up quarks and a down quark, or for simplicity, u u d. The total charge of the quarks adds up to $+e$.
- A neutron has one up quark and two down quarks, or u d d. The total charge of the quarks adds up to zero, so a neutron is uncharged.

24.5 Beta decay

Radioactivity is the decay of unstable nuclei. There is more detail on this in Topic 25.1, Radioactivity. Some nuclei decay by emitting beta radiation. There are two types of **beta decays**; beta-minus (β^-) and beta-plus (β^+). The weak nuclear force is responsible for this type of decay.

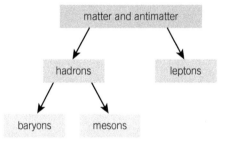

▲ **Figure 1** *Classification of particles*

Transformation of quarks

The two types of beta decays can be summarised by the following decay equations. The **neutrinos** have negligible mass and have no charge. Notice how the nucleon number and the proton number (and therefore charge) are conserved.

Beta-minus (β^-) decay: neutron → proton + electron + electron antineutrino

$$^{1}_{0}n \rightarrow {}^{1}_{1}p + {}^{0}_{-1}e + \overline{\nu}_e$$

At a quark level, one of the down quarks changes into an up quark:

$$u\,d\,d \rightarrow u\,u\,d + {}^{0}_{-1}e + \overline{\nu}_e \qquad \text{or} \qquad d \rightarrow u + {}^{0}_{-1}e + \overline{\nu}_e$$

Beta-plus (β^+) decay: proton → neutron + positron + electron neutrino

$$^{1}_{1}p \rightarrow {}^{1}_{0}n + {}^{0}_{+1}e + \nu_e$$

At a quark level, one of the up quarks changes into a down quark:

$$u\,u\,d \rightarrow u\,d\,d + {}^{0}_{+1}e + \nu_e \qquad \text{or} \qquad u \rightarrow d + {}^{0}_{+1}e + \nu_e$$

Summary questions

1. Explain what is meant by fundamental particles. Give two examples. *(2 marks)*
2. State the force experienced by quarks within a stable hadron. *(1 mark)*
3. Write the decay equation for a neutron decaying into a proton and state two quantities conserved. *(3 marks)*
4. Describe the nature of the gravitational force and the strong nuclear force acting between two protons. *(2 marks)*
5. Use Table 2 to show that the charge on a neutron is zero. *(2 marks)*
6. Complete the reaction: $u\,u\,d \rightarrow ?\,?\,d + {}^{0}_{+1}e + ?$ *(2 marks)*
7. Suggest how we know that there must be more matter than antimatter in the universe. *(2 marks)*
8. Calculate the following ratio for two protons inside a nucleus: $\dfrac{\text{gravitational force on a proton}}{\text{electrostatic force on a proton}}$. *(4 marks)*

Chapter 24 Practice questions

1. How many neutrons are there in the nucleus of $^{7}_{4}$Li?
   - A 3
   - B 4
   - C 7
   - D 10 *(1 mark)*

2. Which statement is correct about beta-minus decay?
   - A A proton transforms into a neutron.
   - B A neutron transforms into a proton.
   - C A neutron transforms into an up quark.
   - D A proton transforms into a down quark. *(1 mark)*

3. In the alpha-scattering experiment, most of the alpha particles fired into a foil went straight though without any deflection.
   What can be deduced from this observation?
   - A Gold atoms are mainly vacuum.
   - B Alpha particles are small particles.
   - C Alpha particles have positive charge.
   - D Gold nucleus is surrounded by electrons. *(1 mark)*

4. The nucleus of $^{16}_{8}$O has more nucleons than the nucleus of $^{4}_{2}$He.
   Which statement is correct about the density of these two nuclei?
   The density of the oxygen nucleus is:
   - A greater than the helium nucleus
   - B smaller than the helium nucleus
   - C the same as that of the helium nucleus
   - D four times that of the helium nucleus. *(1 mark)*

5. Figure 1 shows a graph of radius R of a nucleus and its nucleon number A.
   a Use the graph to confirm the relationship between R and A. *(2 marks)*
   b The mass of the proton and the neutron is about the same.
   Explain how the mass, volume, and density of a nucleus depends on its nucleon number A. *(3 marks)*
   c Use your answer to **a** to estimate the density of the nucleus $^{4}_{2}$He.
   The mass of this nucleus is 4.00 u. *(4 marks)*

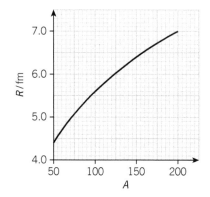

▲ Figure 1

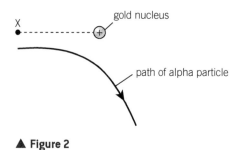

▲ Figure 2

6. Alpha particles, each with kinetic energy 8.2×10^{-13} J, are fired into a thin gold foil. The atomic number of gold is 79.
   a Figure 2 shows the path of one alpha particle as it travels past a gold nucleus.
   Explain why the path shows greater curvature when the alpha particle is closer to the gold nucleus. *(1 mark)*
   b Show that the closest distance an alpha particle can get to the gold nucleus is about 4.4×10^{-14} m. *(3 marks)*
   c Calculate the electrostatic force acting on the gold nucleus when the alpha particle is 4.4×10^{-14} m from the gold nucleus. *(2 marks)*

Chapter 25 Radioactivity

In this chapter you will learn about ...

- ☐ Radioactivity
- ☐ Alpha, beta, and gamma radiations
- ☐ Nuclear transformation equations
- ☐ Activity
- ☐ Half-life
- ☐ Decay constant
- ☐ Exponential decay equation
- ☐ Modelling radioactive decay using spreadsheets
- ☐ Radioactive dating

25 RADIOACTIVITY
25.1 Radioactivity
25.2 Nuclear decay equations
Specification reference: 6.4.3

25.1 Radioactivity

Most nuclei of atoms are stable but some are unstable and consequently emit nuclear radiations in the form of particles or high-energy gamma photons. All nuclear radiations are described as **ionising radiations** because they can ionise atoms by removing some of their electrons, leaving behind positive ions.

There are four different types are radiation: alpha (α), beta-minus (β^-), beta-plus (β^+), and gamma (γ).

Properties and characteristics of the radiations
Alpha
- Each alpha particle is a helium nucleus (two proton and two neutrons).
- An alpha particle has mass 4.00151 u and a charge $+2e$ – this makes it very ionising.
- Alpha particles have a very short range in air and are easily stopped by a thin sheet of paper.

Beta-minus
- Each beta-minus particle is an electron of mass 0.00055 u and a charge $-e$.
- Beta-minus particles are less ionising than alpha particles.
- Beta-minus particles can be stopped by about 1–3 mm of aluminium.

Beta-plus
- Each beta-plus particle is a positron of mass 0.00055 u and a charge $+e$.
- Positrons get annihilated by electrons, and therefore cannot travel too far.

Gamma
- Gamma photons have no charge and travel at the speed of light.
- Gamma photons are poor ionisers.
- Most of the gamma photons can be stopped by lead of thickness of a few cm.

> **Revision tip**
> All types of radiation cause ionisation and therefore can damage living cells.

> **Revision tip**
> Alpha and beta particles are affected by electric and magnetic fields because they are charged particles.

25.2 Nuclear decay equations

The nucleon number and the proton number (and therefore charge) are conserved in all nuclear reactions and decays. The other quantities conserved are mass–energy (see Topic 26.1, Einstein's mass–energy equation) and momentum.

The original nucleus before the decay is known as the **parent nucleus** and the transformed nucleus after the decay is known as the **daughter nucleus**.

Alpha decay
- An unstable nucleus with more than 82 protons is likely to decay by emitting an alpha particle.
- A helium nucleus is emitted from the parent nucleus.
- The daughter nucleus is of a different element.
- Decay equation: $^{A}_{Z}X \rightarrow \ ^{A-4}_{Z-2}Y + \ ^{4}_{2}He$

Radioactivity

Beta-minus decay
- An unstable nucleus with 'too many' neutrons (neutron-rich) is likely to decay by emitting an electron.
- A neutron inside the parent nucleus changes into a proton, electron, and an electron antineutrino.
- Decay equation: $^A_Z X \rightarrow ^A_{Z+1} Y + ^0_{-1} e + \overline{\nu}_e$

Beta-plus decay
- An unstable nucleus with 'too many' protons (proton-rich) is likely to decay by emitting a positron.
- A proton inside the parent nucleus changes into a neutron, positron, and an electron neutrino.
- Decay equation: $^A_Z X \rightarrow ^A_{Z-1} Y + ^0_{+1} e + \nu_e$

Gamma decay
- Surplus energy within the nucleus is released as a gamma photon.
- Decay equation: $^A_Z X \rightarrow ^A_Z X + \gamma$

▼ **Table 1** Changes in nucleon and proton numbers

Decay	Effect on A	Effect on Z
α	Decreases by 4	Decreases by 2
β⁻	No change	Increases by 1
β⁺	No change	Decreases by 1
γ	No change	No change

Table 1 summaries the changes in the nucleon number A and proton number Z as a parent nucleus decays into a daughter nucleus.

Worked example: A decay chain

A nucleus of uranium-238 ($^{238}_{92}$U) eventually transforms into a stable nucleus of lead (Pb) after emitting 8 alpha particles and 6 beta-minus particles. Predict this isotope of lead.

Step 1: Calculate the changes in A and Z due to the alpha-emissions.

decrease in $A = 4 \times 8 = 32$ decrease in $Z = 2 \times 8 = 16$

Step 2: Calculate the changes in A and Z due to the beta-emissions.

change in $A = 0$ increase in $Z = 1 \times 6 = 6$

Step 3: Determine the A and Z values of the lead isotope.

$A = 238 - 32 = 206$ $Z = 92 - 16 + 6 = 82$

The isotope of lead is $^{206}_{82}$Pb.

Summary questions

1. The nucleus of magnesium-28 decays as follows: $^{28}_{12}$Mg $\rightarrow$ $^{28}_{13}$Al + $^0_{-1}$e + $\overline{\nu}_e$
   Identify the particles $^0_{-1}$e and $\overline{\nu}_e$. *(2 marks)*

2. For the decay shown in **Q1**:
   a. state two numbers conserved; *(1 mark)*
   b. calculate the number of neutrons in the nucleus of $^{28}_{12}$Mg and in the nucleus of $^{28}_{13}$Al. *(2 marks)*

3. Complete the following decay equations:
   a. $^{204}_{82}$Pb $\rightarrow$ $^?_?$Hg + $^4_2$He *(1 mark)*
   b. $^?_?$Cf $\rightarrow$ $^{245}_{96}$Cm + $^4_2$He *(1 mark)*

4. Complete the following decay equations:
   a. $^{19}_8$O $\rightarrow$ $^?_?$F + $^0_{-1}$e + $\overline{\nu}_e$ *(1 mark)*
   b. $^{21}_?$Na $\rightarrow$ $^?_{10}$Ne + $^0_{+1}$e + ν_e *(2 marks)*

5. The count rate from a gamma source in air obeys an inverse square law with distance. A Geiger-Müller tube with an opening of area 2.0×10^{-4} m² placed at a distance of 30 cm detects 120 gamma ray photons (counts). Estimate the number of gamma photons emitted by the source per second. *(3 marks)*

6. A single alpha particle produces about 10^4 ions per mm in air. It takes about 10 eV to ionise an atom. Estimate the speed of an alpha particle that has a range of 3.0 cm in air.
   mass of alpha particle = 6.6×10^{-27} kg *(4 marks)*

25.3 Half-life and activity
25.4 Radioactive decay calculations

Specification reference: 6.4.3

25.3 Half-life and activity

Natural **radioactivity** is the *random* and *spontaneous* decay of unstable nuclei. The decay is random because we cannot predict when a particular nucleus in a substance will decay or which one will decay next – each nucleus within a substance has the same chance of decaying per unit time. The decay is spontaneous because the decay of nuclei is not affected by the presence of other nuclei in the substance or external factors (e.g., pressure).

Half-life, activity, and decay constant

The **half-life** $t_{\frac{1}{2}}$ of an isotope is the average time it takes for half the number of active nuclei in the sample to decay.

Even isotopes of the same element can have a wide range of half-lives. This is illustrated in Table 1 for some of the isotopes of carbon.

▼ **Table 1** *Half-lives of some carbon isotopes*

Isotope	$^{9}_{6}C$	$^{10}_{6}C$	$^{11}_{6}C$	$^{12}_{6}C$	$^{13}_{6}C$	$^{14}_{6}C$	$^{15}_{6}C$
Half-life	0.126 s	19.3 s	20.3 min	stable	stable	5730 y	2.45 s

The **activity** A of a substance is the rate at which the nuclei decay (disintegrate).

The SI unit of **activity** is the **becquerel** (Bq). 1 Bq = 1 decay per second.

What is meant by a beta-emitting sample having an activity of 2000 Bq? You could imagine 2000 nuclei in the sample decaying per second or 2000 beta-particles emitted per second.

Consider a source with a known isotope and where the parent nuclei decay into stable daughter nuclei. The activity A of a source is directly proportional to the number of undecayed nuclei N left in the source. The activity is given by the equation

$$A = \lambda N$$

where λ is the decay constant of the isotope. The unit for decay constant is normally s^{-1}, but min^{-1}, h^{-1}, and y^{-1} are frequently used.

The **decay constant** can be defined as the probability of decay of an individual nucleus per unit time.

25.4 Radioactive decay calculations

The decay of a radioactive substance follows an exponential pattern similar to the discharge of a capacitor. The rate of decay of a radioactive substance depends on the half-life and the decay constant of the isotope.

Exponential decay equation

The number of undecayed nuclei N left in the source at time t is given by the equation

$$N = N_0 e^{-\lambda t}$$

where N_0 is the number of undecayed nuclei at time $t = 0$ and e is the base of natural logarithms, 2.718. The number of undecayed nuclei decreases exponentially with time, see Figure 1.

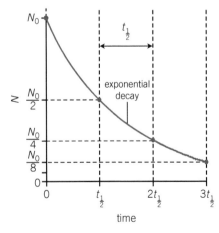

▲ **Figure 1** *A graph of number of undecayed nuclei N against time showing exponential decay*

Radioactivity

The activity A of the source is directly proportional to N. Therefore
$$A = A_0 e^{-\lambda t}$$
where A_0 is the activity at time $t = 0$.

The decay constant λ and half-life $t_{\frac{1}{2}}$ are inversely proportional to each other and are given by the equation
$$\lambda t_{\frac{1}{2}} = \ln(2)$$

The half-life of the carbon-14 isotope is 5730 y. You can show that its decay constant is $1.21 \times 10^{-4}\,\text{y}^{-1}$ or $3.83 \times 10^{-12}\,\text{s}^{-1}$.

Revision tip
Instead of the equation $N = N_0 e^{-\lambda t}$ you can use $N = (0.5)^n \times N_0$ where n is the number of half-lives elapsed.

Worked example: Activity

A sample of a radioactive substance has 6.0×10^{10} nuclei of an isotope at time $t = 0$. The decay constant of the isotope is $2.0 \times 10^{-3}\,\text{s}^{-1}$. Calculate the activity at time $t = 6.0$ minutes.

Step 1: Write down the information given.

$N_0 = 6.0 \times 10^{10}$ $\lambda = 2.0 \times 10^{-3}\,\text{s}^{-1}$ $t = 6.0 \times 60 = 360\,\text{s}$ $A = ?$

Step 2: Calculate the initial activity A_0.

$A_0 = \lambda N_0 = 2.0 \times 10^{-3} \times 6.0 \times 10^{10} = 1.2 \times 10^8\,\text{Bq}$

Step 3: Calculate the activity at time $t = 360\,\text{s}$.

$A = A_0 e^{-\lambda t} = 1.2 \times 10^8 \times e^{-2.0 \times 10^{-3} \times 360} = 5.8 \times 10^7\,\text{Bq}$ (2 s.f.)

Summary questions

1. What is the relationship between decay constant and half-life? *(1 mark)*
2. An alpha-emitting source has an activity of 120 Bq. Estimate the number of alpha particles emitted in a period of 2.0 s and state any assumption made. *(2 marks)*
3. A source has an isotope of half-life 2.0 minutes. The sample has 8.0×10^{15} undecayed nuclei of the isotope. Calculate:
   a. the number of nuclei left in the source after 4.0 minutes; *(2 marks)*
   b. the total number of nuclei decayed in the source after 6.0 minutes. *(3 marks)*
4. The activity of an alpha-emitting source is 3.4×10^{10} Bq. The kinetic energy of each alpha particle is 1.5 MeV. Calculate the power of the source. *(3 marks)*
5. The isotope of thorium-234 has a half-life of 6.7 h. Calculate the activity of a sample of thorium of mass 1.5 mg. (The molar mass of thorium-234 is 0.234 kg mol^{-1}.) *(4 marks)*
6. Use the information given in Table 1 to calculate the time in years it takes for the activity of a small sample of carbon-14 to decrease to 72 % of its initial activity. *(4 marks)*

25.5 Modelling radioactive decay
25.6 Radioactive dating

Specification reference: 6.4.3

25.5 Modelling radioactive decay

The activity A of a source is given by the equation $A = \lambda N$, where λ is the decay constant of the isotope and N is the number of undecayed nuclei of the isotope in the source. This equation can be mathematically written as

$$\frac{\Delta N}{\Delta t} = -\lambda N$$

where the activity A, which is the rate of decay of nuclei, is written as $\frac{\Delta N}{\Delta t}$ and the minus sign is necessary because the number of active nuclei in the source decreases with time t. You have already met the 'solution' of the rate of decay equation, $N = N_0 e^{-\lambda t}$.

You can also use a spreadsheet method to estimate the number of nuclei left in the source.

Using spreadsheets

Imagine there are N nuclei in a sample. The number of nuclei decaying in a very small interval of time Δt is given by $\Delta N = (\lambda \Delta t)N$. The number of nuclei left in the sample after Δt will be ΔN subtracted from the initial number. This process can be repeated to determine the number of nuclei N at time t.

This is easily illustrated in the example below.

$N_0 = 6.00 \times 10^{10}$ nuclei $t_{\frac{1}{2}} = 30$ s $\lambda = \frac{\ln(2)}{30} = 0.0231$ s^{-1} $\Delta t = 1.00$ s

$$\Delta N = (\lambda \Delta t) \times N = (0.0231 \times 1.00) N$$

or

$$\Delta N = 0.0231 N$$

This means that 2.31 % of the nuclei have decayed leaving 97.69 % of the previous number of nuclei.

time /s	0.00	1.00	2.00	3.00	4.00	5.00	6.00	
$N / 10^{10}$	6.00	5.86	5.72	5.59	5.46	5.34	5.21	....etc.

$N = 0.9769 \times 6.00 \times 10^{10}$

$N = 0.9769^2 \times 6.00 \times 10^{10}$

The number of nuclei N left in the sample can be determined accurately by making the interval Δt even smaller, and this is where a spreadsheet is helpful.

25.6 Radioactive dating

The isotope of carbon-14 has a half-life of 5730 years. The isotopes carbon-12 (stable) and carbon-14 are used in the technique of **carbon-dating** to determine the age of organic materials. The actual number of carbon-14 and carbon-12 nuclei in a small sample of organic material can be determined accurately using mass spectrometers

Carbon-dating

The ratio of carbon-14 nuclei to carbon-12 nuclei in atmospheric carbon is almost constant at about 1.3×10^{-12}.

The ratio is the same in all living (organic) things. Once the living thing (e.g., a tree or a person) dies, it stops taking in carbon. The carbon-14 nuclei within the dead material decreases exponentially with time. The age of the organic material can be determined by comparing ratios of carbon-14 nuclei to carbon-12 nuclei of the dead material and a similar living material. See the worked example below.

The ratio of carbon-14 nuclei to carbon-12 nuclei cannot be assumed to be constant over time. This ratio is affected by natural events such as volcanic eruptions and increased carbon-emission from cars and burning fossil fuels. The value of this ratio in the past is not known. This introduces a large uncertainty in the age of a relic.

> **Revision tip**
> The isotope of rubidium-87, with half-life of 49 billion years, is used to determine the age of ancient rocks on the Earth.

 Worked example: Age of a relic

A mass spectrometer is used to determine the ratio r of carbon-14 nuclei to carbon-12 nuclei in a living tree and a wooden axe found in a cave. Use the information below to determine the age of the axe.

$r = 1.3 \times 10^{-12}$ for the living tree $r = 7.2 \times 10^{-13}$ for the axe

Step 1: Determine the percentage of carbon-14 nuclei left in the axe.

$$\% \text{ of carbon-14 left in the axe} = \frac{7.2 \times 10^{-13}}{1.3 \times 10^{-12}} \times 100 = 55.4\%$$

Step 2: Use $N = N_0 e^{-\lambda t}$ to determine the time t.

$$\lambda = \frac{\ln(2)}{5730} = 1.21 \times 10^{-4} \, \text{y}^{-1}$$

$$\frac{N}{N_0} = 0.554 = e^{-\lambda t}$$

$$t = -\frac{\ln(0.554)}{1.21 \times 10^{-4}} = 4880 \, \text{y} = 4900 \, \text{y} \, (2 \text{ s.f.})$$

Summary questions

1. A source contains an isotope of half-life 10 s. The source has 1000 nuclei of this isotope.
   Suggest why a time interval Δt of 5.0 s is not suitable when using the equation $\Delta N = (\lambda \Delta t)N$. *(1 mark)*
2. **a** Calculate the decay constant of the isotope in **Q1**. *(1 mark)*
   b Estimate the number of nuclei decaying in an interval of 0.10 s. *(2 marks)*
3. A sample of ancient wood has an activity of 0.10 Bq. A sample of living wood of the same mass has an activity of 0.40 Bq. Calculate the age of the ancient wood. *(2 marks)*
4. The activity of a living wood from the decay of carbon-14 nuclei is 1.7 Bq.
   Calculate the number of carbon-14 nuclei in this living wood. *(3 marks)*
5. A certain sample of dead wood is found to have an activity of 0.32 Bq. An identical mass of living wood has an activity of 1.6 Bq. Calculate the age of the dead wood. *(4 marks)*
6. The isotope of rubidium-87 has a half-life of 49 billion years. A sample of rock on the Earth is found to have 94% of its rubidium-87 nuclei left. Estimate the age of the Earth. *(4 marks)*

Chapter 25 Practice questions

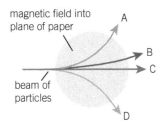

▲ Figure 1

1 A beam of radiation from a mixture of radioactive substances is passed through a uniform magnetic field, see Figure 1.
  Which is the correct path described by the alpha particles? *(1 mark)*

2 The table below shows the number of nuclei N in the substance and the decay constant λ of the isotope of four substances A, B, C, and D.

Substance	A	B	C	D
$N / 10^{16}$	2.0	4.0	6.0	8.0
$\lambda / 10^{-6} \text{s}^{-1}$	7.0	5.0	3.0	1.0

  Which substance has the greatest activity? *(1 mark)*

3 The nucleus of $^{233}_{91}\text{Pa}$ emits a beta-minus particle and then an alpha particle. It transforms into a nucleus of thorium Th.
  Which is the correct transformed isotope of thorium?
  A $^{228}_{89}\text{Th}$
  B $^{229}_{89}\text{Th}$
  C $^{229}_{90}\text{Th}$
  D $^{233}_{90}\text{Th}$ *(1 mark)*

4 The activity of a radioactive substance is 700 Bq.
  What is the activity of the substance after 3.5 half-lives?
  A 57 Bq
  B 62 Bq
  C 88 Bq
  D 200 Bq *(1 mark)*

5 This question is about the radioactive decay of carbon-10 nuclei in a source.
  Figure 2 shows the activity A of the source from time $t = 30$ s to about $t = 45$ s.

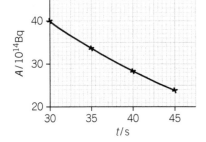

▲ Figure 2

  a Use Figure 2 to show that the activity decreases exponentially with time. *(2 marks)*
  b Use Figure 2 to show that the half-life of carbon-10 ($^{10}_{6}\text{C}$) is about 20 s. *(4 marks)*
  c Carbon-10 is a beta-plus emitter.
     Write the decay equation for the carbon-10 nucleus. *(2 marks)*
  d The kinetic energy of each beta-plus particle is 3.4×10^{-13} J.
     Calculate the power of the source at time $t = 30$ s. *(2 marks)*

6 **a** A student uses a Geiger-Müller tube and a counter to record the count rate from a radioactive source. Another student suggests the 'count rate is the activity of the source'. *(2 marks)*
     Discuss whether this suggestion is correct or not.
  b A sample of a radioactive isotope contains 2.0×10^9 active nuclei.
     The half-life of the isotope is 6.0 hours. Calculate:
     i the initial activity of the sample; *(3 marks)*
     ii the number of active nuclei of the isotope remaining after 12 hours; *(2 marks)*
     iii the number of active nuclei of the isotope remaining after 21 hours. *(2 marks)*

Chapter 26 Nuclear physics

In this chapter you will learn about ...

- ☐ Einstein's mass–energy equation
- ☐ Radioactivity
- ☐ Annihilation
- ☐ Creation of matter
- ☐ Binding energy
- ☐ Mass defect
- ☐ Binding energy per nucleon
- ☐ Nuclear fission
- ☐ Chain reaction
- ☐ Fission reactors
- ☐ Nuclear fusion

26 NUCLEAR PHYSICS
26.1 Einstein's mass–energy equation
26.2 Binding energy

Specification reference: 6.4.4

26.1 Einstein's mass–energy equation

Einstein's mass–energy equation is $\Delta E = \Delta mc^2$, where ΔE is the change in energy, Δm is the change in mass, and c is the speed of light in a vacuum. Mass and energy are equivalent. In nuclear reactions, it is 'mass–energy' that is conserved and not just 'energy'.

Two important outcomes for a system:

- Mass increases when energy increases.
- Mass decreases when energy decreases.

When you are travelling in a fast car, your mass is greater than your mass at rest (**rest mass**). The change in mass is incredibly small and therefore not noticeable. However, the mass of an electron in a particle accelerator can increase by a substantial amount.

Radioactivity

In natural radioactivity where does the energy of the emitted particles and photons come from?

The energy comes from matter being converted into energy. In all the nuclear decays below, the mass of the parent nucleus is *greater* than the total mass of the 'product' particles.

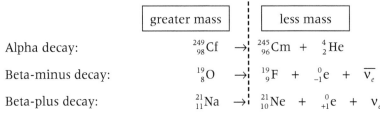

| greater mass | less mass |

Alpha decay: $^{249}_{98}\text{Cf} \rightarrow \, ^{245}_{96}\text{Cm} + \, ^{4}_{2}\text{He}$

Beta-minus decay: $^{19}_{8}\text{O} \rightarrow \, ^{19}_{9}\text{F} + \, ^{0}_{-1}\text{e} + \overline{\nu}_e$

Beta-plus decay: $^{21}_{11}\text{Na} \rightarrow \, ^{21}_{10}\text{Ne} + \, ^{0}_{+1}\text{e} + \nu_e$

Matter annihilation and creation

Energy can be created from matter. When a particle and its corresponding antiparticle meet, they completely destroy each other and their entire mass is converted into two identical gamma photons. This process is called **annihilation**, see Figure 1a. The energy of each photon is mc^2, where m is the mass of the particle or the antiparticle. Electron–positron annihilation is used in PET scanners.

Matter can be created from energy. A single gamma photon of energy is equal to or greater than $2mc^2$ can produce a particle and an antiparticle, each of mass m. This process is known as **pair production.** See Figure 1b.

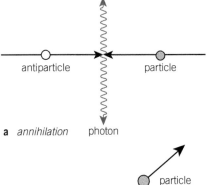

▲ **Figure 1** (a) Annihilation. (b) Pair production

Synoptic link

You will learn more about PET scanners in Topic 27.5, PET scans.

Worked example: Matter creation

Calculate the minimum energy in MeV of a photon capable of creating an electron–positron pair.

Step 1: Write down the known quantities.

mass of electron = mass of positron = 9.11×10^{-31} kg $c = 3.00 \times 10^8 \, \text{m s}^{-1}$

Step 2: Use Einstein's mass–energy equation to calculate the energy of the photon.

$\Delta E = \Delta mc^2 = 2 \times 9.11 \times 10^{-31} \times (3.00 \times 10^8)^2 = 1.64 \times 10^{-13}$ J

Step 3: Convert the energy into MeV. 1 eV = 1.60 × 10⁻¹⁹ J.

$$\text{energy of photon} = \frac{1.64 \times 10^{-13}}{1.60 \times 10^{-19}} = 1.02 \times 10^6 \text{ eV}$$

energy of photon = 1.02 MeV (3 s.f.)

26.2 Binding energy

The mass of a nucleus is always *less* than the total mass of its separate nucleons (neutrons and protons). The nucleons are held together within the nucleus by the strong nuclear force. According to the mass–energy equation, external energy is required to pull apart these nucleons.

Mass defect and binding energy

Binding energy of a nucleus is related to its mass defect.

The **binding energy** of a nucleus is defined as the minimum energy required to completely separate a nucleus into its constituent protons and neutrons.

The **mass defect** of a nucleus is defined as the difference between the mass of the completely separated nucleons and the mass of the nucleus.

To calculate the binding energy of a nucleus, you can use the mass–energy equation.

$$\text{binding energy of nucleus} = \text{mass defect of nucleus} \times c^2$$

Binding energy per nucleon

The average **binding energy per nucleon** is useful when comparing different nuclei. The greater the binding energy per nucleon, the more tightly bound are the nucleons within the nucleus. Figure 2 shows the binding energy (BE) per nucleon against nucleon number A for nuclei.

- The isotope of iron-56 has the greatest BE per nucleon – its nucleons are the most tightly bound together.
- In natural radioactive decay, the BE of the product particles is greater than the BE of the parent nucleus. The difference in the binding energies is the energy released.
- BE increases in fission and fusion processes and therefore energy is released. There is more detail in Topic 26.3, Nuclear fission, and Topic 26.4, Nuclear fusion.

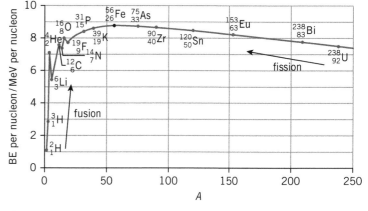

▲ **Figure 2** *BE per nucleon against nucleon number graph for nuclei*

Summary questions

1. Convert a mass of 1.0 mg into energy in joules. *(2 marks)*
2. Convert the mass of an electron into energy in joules. *(2 marks)*
3. State and explain the change in the mass of the following systems:
   a. A lump of hot iron cooling in a room. *(2 marks)*
   b. An electron slowing down. *(2 marks)*
   c. A proton getting faster in an accelerator. *(2 marks)*
4. Use Figure 2 to estimate the binding energy of the following nuclei in MeV.
   a. $^{2}_{1}\text{H}$ b. $^{4}_{2}\text{He}$ c. $^{238}_{92}\text{U}$ *(6 marks)*
5. The mass of the $^{16}_{8}\text{O}$ nucleus is 2.656×10^{-26} kg. Calculate its BE per nucleon in MeV. *(4 marks)*
6. There is a decrease in mass of 5.8×10^{-3} u when a polonium-210 emits an alpha particle and transforms to an isotope of lead. Calculate the total energy released by 1 mol (0.210 kg) of polonium-210. *(4 marks)*

26.3 Nuclear fission
26.4 Nuclear fusion

Specification reference: 6.4.4

26.3 Nuclear fission

Nuclear power stations use **fission** reactions to generate power. The energy released in fission reactions can be as much as a million times greater than burning a similar mass of fossil fuels.

Induced fission

Induced fission occurs when a nucleus absorbs a neutron and then splits into two approximately equal fragment nuclei and a few fast-moving neutrons.

A typical fission reaction of a uranium-235 nucleus is shown below:

$$^1_0n + ^{235}_{92}U \rightarrow ^{236}_{92}U^* \rightarrow ^{141}_{56}Ba + ^{92}_{36}Kr + 3^1_0n$$

A slow-moving neutron (**thermal neutron**) is absorbed by the uranium-235 nucleus. It temporarily forms the highly unstable nucleus of uranium-236 which very quickly splits into smaller nuclei of barium-141 and krypton-92 and also three fast-moving neutrons. In this single reaction about 180 MeV of energy is released as kinetic energy of the fragment nuclei and neutrons.

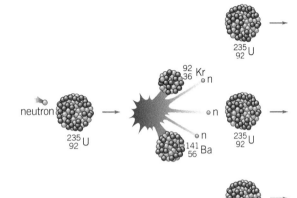

▲ **Figure 1** *Chain reaction*

- The chance of a fission reaction for a uranium-235 nucleus is greater with slow-moving neutrons than fast-moving neutrons.
- The total mass of the particles after the fission reaction is always less than the total mass of the particles before the reaction. The difference in the mass Δm is released as kinetic energy equal to Δmc^2.
- The total binding energy of the particles after fission is greater than the total binding energy before it. The difference in the binding energies is equal to the energy released.
- In a **chain reaction**, the neutrons released in a fission reaction can trigger further fission reactions, see Figure 1.

Fission reactor

The main components of a fission reactor are: fuel rods, control rods and moderator. See Figure 2.

- The fuel rods contain enriched uranium (uranium-238 with 2–3% uranium-235).
- The nuclei of the material of the **moderator** slow down the fast neutrons produced in the fission reactions without absorbing them. Water and graphite (carbon) are the two commonly used materials.
- The nuclei of the material of the **control rods** easily absorb neutrons. The control rods can be moved in and out of the reactor core to control the rate of the fission reactions. Boron and cadmium are the two commonly used materials.

Storage and disposal of the by-products from fission reactors is difficult because of the toxicity and the long half-lives of many of the isotopes.

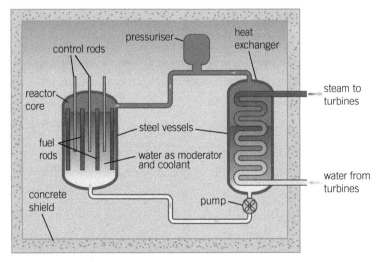

▲ **Figure 2** *The main components of a nuclear reactor. The water is a coolant; it removes thermal energy from the reactor.*

Nuclear physics

26.4 Nuclear fusion

Fusion is the process by which stars produce energy. The core of a star provides the high temperature (10^8 K) and high pressure necessary for the fusion of hydrogen nuclei into helium nuclei.

Fusion

Fusion occurs when two high-speed nuclei combine to form a bigger nucleus.

The positive nuclei repel each other. The nuclei therefore need to have large enough kinetic energy for them to get close enough (< 3 fm) for the strong nuclear force to bind them together.

A fusion reaction is shown below:

$$^2_1\text{H} + ^1_1\text{p} \rightarrow ^3_2\text{He}$$

- The mass of the $^3_2$He nucleus is less than the total mass the $^2_1$H nucleus and the proton $^1_1$p. The difference in the mass Δm is released as energy equal to Δmc^2.
- The binding energy of the $^3_2$He nucleus is greater than the binding energy of the $^2_1$H nucleus (the proton is a lone particle and has no BE). The difference in the binding energies is equal to the energy released.

Worked example: Energy from fusion

One of the many fusion reactions taking place in the core of a star is

$$^2_1\text{H} + ^2_1\text{H} \rightarrow ^4_2\text{He}$$

Use the binding energy (BE) per nucleon against nucleon number graph in Topic 26.2, Binding energy, to estimate the energy in joules released in this reaction.

Step 1: Write down the BE per nucleon of the nuclei.

BE per nucleon of $^2_1$H = 1.0 MeV   BE per nucleon of $^4_2$He = 7.1 MeV

Step 2: Calculate the BE of each nucleus.

BE of $^2_1$H = 2 × 1.0 = 2.0 MeV   BE of $^4_2$He = 4 × 7.1 = 28 MeV

Step 3: The energy released is the difference in the BE. Calculate this energy.

energy released = 28.4 − 2.0 = 26.4 MeV

energy released = 26.4 × 1.60 × 10^{-19} (1 eV = 1.60 × 10^{-19} J)

energy released = 4.2 × 10^{-12} J (2 s.f.)

Summary questions

1. State one similarity between fission and fusion reactions. *(1 mark)*
2. Show that the nucleon and the proton numbers are conserved in each reaction below.
   a $^2_1\text{H} + ^1_1\text{p} \rightarrow ^3_2\text{He}$ *(2 marks)*
   b $^1_0\text{n} + ^{235}_{92}\text{U} \rightarrow ^{141}_{56}\text{Ba} + ^{92}_{36}\text{Kr} + 3^1_0\text{n}$ *(2 marks)*
3. Explain why nuclei of hydrogen will only fuse together at high temperatures. *(2 marks)*
4. A single fission reaction of uranium-235 can produce energy of about 2.0 × 10^{-11} J. Estimate the energy produced from 1.0 kg of uranium-235.
   molar mass of uranium-235 = 0.235 kg mol^{-1} *(3 marks)*
5. Use the binding energy (BE) per nucleon against nucleon number graph in Topic 26.2, Binding energy, to estimate the energy in joules released in the fusion reaction $^1_1\text{p} + ^1_1\text{p} \rightarrow ^2_1\text{H} + ^0_{+1}\text{e} + \nu$. *(3 marks)*
6. Use your answer to Q5 to estimate the energy produced from 1.0 kg of hydrogen-1.
   molar mass of hydrogen-1 = 1.0 g mol^{-1} *(3 marks)*

Chapter 26 Practice questions

1 Annihilation occurs when:
   A electron and positron meet
   B electron and proton interact
   C electron and a photon collide
   D electron and neutrino interact *(1 mark)*

2 In a fusion reaction, hydrogen nuclei $^{2}_{1}H$ and $^{1}_{1}H$ combine together to form a nucleus of helium $^{3}_{2}He$. These nuclei have binding energy BE.
   The BE of $^{3}_{2}He$ nucleus is _____ the total BE of the hydrogen nuclei. What is the missing word or words?
   A twice
   B equal to
   C less than
   D greater than *(1 mark)*

3 What is the energy equivalent to the mass of an electron?
   A 2.7×10^{-22} J
   B 8.2×10^{-14} J
   C 1.6×10^{-19} J
   D 1.5×10^{-10} J *(1 mark)*

4 The binding energy per nucleon for the nucleus of $^{92}_{36}Kr$ is 8.7 MeV. What is the binding energy of this nucleus?
   A 310 MeV
   B 490 MeV
   C 800 MeV
   D 1100 MeV *(1 mark)*

5 Use Figure 2 in Topic 26.2 to answer this question.
   a Explain why the isotope of iron-56 cannot decay. *(1 mark)*
   b i Use Figure 1 to explain why energy is released in the fusion reaction shown below.
       $$^{2}_{1}H + ^{2}_{1}H \rightarrow ^{4}_{2}He$$ *(3 marks)*
     ii Estimate the energy released when 1 mole of hydrogen-2 fuse together. *(4 marks)*

6 a Explain what is meant by the statement 'mass and energy are equivalent'. *(1 mark)*
   b One of the many fusion reactions in the Sun is shown below:
       $$^{2}_{1}H + ^{1}_{1}H \rightarrow ^{3}_{2}He$$
       i Explain why high temperatures are necessary for this fusion reaction to occur. *(2 marks)*
       ii In terms of the masses of the particles, explain why energy is released. *(2 marks)*
       iii The energy released in the reaction above is about 5.5 MeV. Calculate the change in mass responsible for this energy. *(3 marks)*
   c Explain the difference between fission and fusion reactions. *(2 marks)*

Chapter 27 Medical imaging

In this chapter you will learn about ...

- ☐ X-rays
- ☐ Interaction of X-rays with matter
- ☐ Intensity of transmitted X-rays
- ☐ Contrast materials
- ☐ CAT scanner
- ☐ Gamma camera
- ☐ PET scanner
- ☐ Ultrasound
- ☐ A and B scans
- ☐ Acoustic impedance
- ☐ Doppler imaging
- ☐ Speed of blood

27 MEDICAL IMAGING
27.1 X-rays
27.2 Interaction of X-rays with matter

Specification reference: 6.5.1

27.1 X-rays

X-rays are electromagnetic waves with a wavelength in the range 10^{-8} m to 10^{-13} m. An X-ray photon has greater energy than a photon of visible light because of its shorter wavelength. X-ray photons are produced when fast-moving electrons are decelerated by the atoms of a metal. The kinetic energy of the electrons is transformed into X-ray photons.

Production of X-rays

Figure 1 shows the main components of an **X-ray tube**. Electrons are accelerated through a high potential difference, typically 100 kV. They travel from the cathode (heater) to the anode (**target metal**). Most of the kinetic energy of the electrons is transformed into heat in the anode. The anode is cooled by circulating water through it. About 1% of the kinetic energy of the electrons is transformed into X-ray photons. These X-ray photons have a range of wavelengths.

▲ **Figure 1** *The main components of an X-ray tube*

27.2 Interaction of X-rays with matter

The intensity of a beam of X-rays decreases as the X-ray photons are either stopped or scattered by the atoms of the material.

Attenuation mechanisms

Table 1 summarises the four **attenuation** mechanisms.

▼ **Table 1**

Attenuation mechanism	Diagram	Energy of photons	Description
Simple scatter	incident X-ray photon, scattered X-ray photon, electron, atom, nucleus	1 to 20 keV	The X-ray photon is scattered elastically by an electron.
Photoelectric effect	incident X-ray photon, removed electron, electron, atom, nucleus	< 0.1 MeV	The X-ray photon disappears and removes an electron from the atom.
Compton scattering	incident X-ray photon, atom, removed electron, electron, nucleus, scattered low-energy photon	0.5 to 5.0 MeV	The X-ray photon is scattered by an electron, its energy is reduced, and the electron is ejected from the atom.
Pair production	incident X-ray photon, electron, atom, nucleus, positron	> 1.02 MeV	The X-ray photon disappears to produce an electron–positron pair.

Medical imaging

Intensity of transmitted X-rays

The **intensity** I of a parallel beam (collimated) of X-rays decreases exponentially with the thickness x of material and is given by the equation

$$I = I_0 e^{-\mu x}$$

where I_0 is the initial intensity and μ is the **attenuation coefficient** or the **absorption coefficient** of the material. The SI unit of μ is m^{-1}, but mm^{-1} and cm^{-1} are also used. Bone has a greater attenuation coefficient than soft tissues – this is why a simple X-ray scan (image) is used to identify broken bones in a patient. Figure 2 shows the variation of intensity I with thickness x of a material.

Contrast materials

Soft tissues and muscles are poor absorbers of X-rays. A contrast medium is a material which is either injected into a patient or ingested by the patient in order to improve the visibility of soft tissues on an X-ray scan. A contrast medium must be harmless to patients.

The main interaction mechanism for the X-rays used in hospitals is the photoelectric effect. For this effect, $\mu \propto Z^3$, where Z is the atomic number of the element. Iodine and barium are the two most frequently used elements within contrast media. Both iodine and barium have larger Z values than soft tissues.

- Barium sulphate (barium meal) is used when imaging the digestive system of a patient.
- Iodine-based liquid is often injected into the blood vessels when diagnosing circulatory problems.

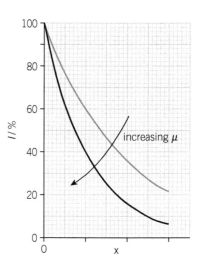

▲ **Figure 2** *The transmitted intensity I decreases exponentially with thickness x of the material*

> **Synoptic link**
>
> Intensity was covered in Topic 11.5, Intensity.

Worked example: Thickness of bone

The attenuation coefficient of bone is $0.60 \, cm^{-1}$. Calculate the thickness of bone that will halve the intensity of X-rays incident on the bone.

Step 1: Write down the quantities given.
$\mu = 0.60 \, cm^{-1}$ $I = 0.50 I_0$

Step 2: Use the equation $I = I_0 e^{-\mu x}$ to calculate the thickness x.

$0.50 I_0 = I_0 e^{-0.60x}$ or $0.50 = e^{-0.60x}$

$\ln(0.50) = -0.60x$

$x = \dfrac{\ln(0.50)}{-0.60} = 1.16 \, cm = 1.2 \, cm$ (2 s.f.)

> **Maths: $\log_e$ or ln**
>
> When $e^x = y$ then $x = \ln(y)$, where ln is an abbreviation for $\log_e$.
>
> It is important that you use the ln button on your calculator and not the lg button (which is $\log_{10}$).

Summary questions

1. Name two attenuation mechanisms where the X-ray photon is scattered. *(1 mark)*
2. An X-ray tube operates using a 100 kV supply. Use Table 1 to state the most dominant attenuation mechanism for the X-ray photon. *(1 mark)*
3. The attenuation coefficient of muscle is about $0.21 \, cm^{-1}$. Calculate the percentage of the original intensity of X-rays transmitted through 3.0 cm of muscle. *(3 marks)*
4. A single X-ray photon is produced from the kinetic energy of a single electron. Calculate the maximum energy of an X-ray photon in J from an X-ray tube connected to a 120 kV supply. *(3 marks)*
5. Calculate the shortest wavelength of an X-ray photon produced from an X-ray tube connected to a 200 kV supply. *(3 marks)*
6. The mean atomic number of elements within soft tissues is 7. The atomic number of barium is 56. Use this information to explain why barium is a useful contrast material. *(3 marks)*

27.3 CAT scans
27.4 The Gamma camera
27.5 PET scans

Specification reference: 6.5.1, 6.5.2

27.3 CAT scans

A computerised axial tomography (CAT) scanner digitally records a large number of cross-sectional scans (images) – very much like a loaf of bread that has been cut into many thin slices. Soft tissues can be distinguished but the patient is exposed to the ionising effects of X-rays for a longer period of time.

CAT scanner

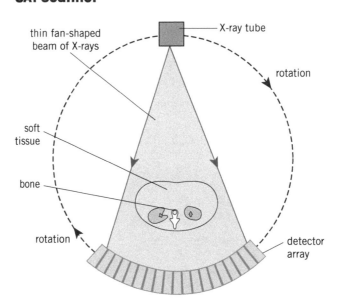

▲ **Figure 1** *The X-ray tube and the detectors can make three revolutions around the patient every second*

- The patient lies on a horizontal table that can move in and out of a large doughnut-shaped vertical ring (gantry).
- The ring contains a single X-ray tube and up to a thousand X-ray detectors on the opposite side. The detectors are all connected to a computer.
- The X-ray tube produces a thin fan-shaped beam of X-rays which irradiate a thin cross-section through the patient, see Figure 1.
- The X-ray tube and the detectors rotate around the patient. A two-dimensional image is digitally recorded by the computer.
- After every revolution, the X-ray tube and the detectors move a short distance (about 1 cm) along the length of the patient, so that in the next revolution, the X-ray beam irradiates the next slice through the patient.
- All the digital images can be manipulated by the computer software to produce cross-sectional images through the patient and also a three-dimensional image of the patient.

> **Revision tip**
> The term 'axial' in CAT refers to the images taken in the axial plane (cross-sections) through the patient. *'Tomos'* is Greek for *'slice'*.

27.4 The Gamma camera

The most common medical tracer contains the isotope of technetium-99m (Tc-99m). The Tc-99m isotope has a half-life of about 6.0 hours and emits 140 keV gamma photons. A compound containing Tc-99m is injected into a patient and this compound will target specific cells in the body. Tc-99m can be used to monitor the function of many major organs (brain, heart, liver, lungs, and kidney). The concentration of Tc-99m within the patient can be used to identify defects in the function of these organs.

Components of the camera

A gamma camera is placed above the patient. Figure 2 shows the main components of the gamma camera.

- **Collimator:** This consists of long and thin cylindrical tubes made from lead. Only gamma photons travelling along the axis of a tube will reach the scintillator.
- **Scintillator:** Each gamma photon hitting the scintillator (sodium iodide) can produce thousands of photons of visible light.
- **Photomultiplier tubes:** Each tube is an electrical device that produces an electrical pulse whenever a photon of visible light is incident on it.
- Computer and display: The electrical pulses from all the photomultiplier tubes are used by the computer and its software to accurately pinpoint the origin of the gamma photons, and therefore the technetium-99m within the patient. An image of the concentration levels of Tc-99m is displayed on a display screen.

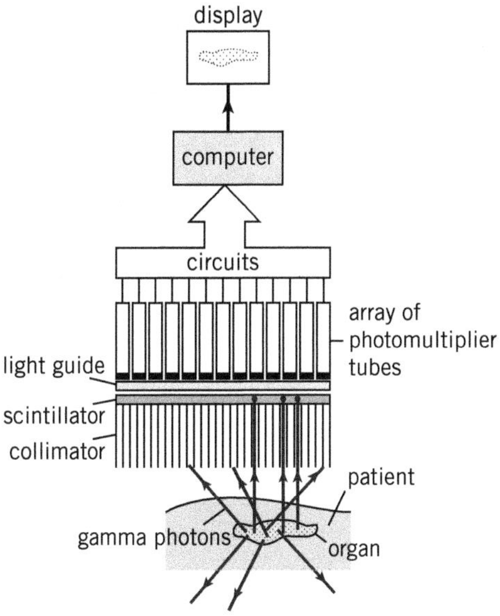

▲ **Figure 2** *Components of a gamma camera*

27.5 PET scans

The most common **medical tracer** used in positron emission tomography (PET) scanning is fluorodeoxyglucose (FDG), which has radioactive atoms of fluorine-18. This isotope of fluorine-18 is a positron emitter with a half-life of about 110 minutes. FDG is injected into a patient and it gathers in tissues with a high rate of respiration. The *function* of organs such as the brain can be observed. PET scanning is an expensive technique because the medical tracers have to be produced on-site.

PET scanner

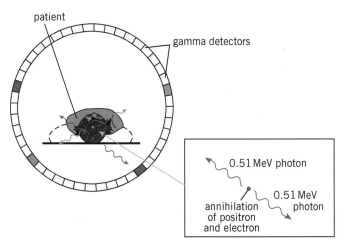

▲ **Figure 3** *The 0.51 MeV gamma photons are detected by the gamma detectors around the patient*

Figure 3 shows a patient surrounded by gamma detectors which are all connected to a computer.

- The positron emitted from a fluorine-18 nucleus does not travel too far before it gets annihilated by an electron.
- The annihilation produces two 0.51 MeV gamma photons emitted in opposite directions.
- The computer software can determine the exact point of annihilation within the patient from the arrival times of these gamma photons at diametrically opposite detectors and the speed of these photons ($3.00 \times 10^8 \, \text{m s}^{-1}$).
- An image of the concentration levels of FDG within the patient can be displayed on a screen.

Summary questions

1. State one advantage of a CAT scan over traditional X-ray image. *(1 mark)*
2. State why a CAT scan can be harmful to a patient. *(1 mark)*
3. Explain why the collimators in a gamma camera need to be made of long and narrow lead tubes. *(2 marks)*
4. Discuss the technique that could be used to assess the effects of a drug on the function of the brain. *(2 marks)*
5. Distances as small as 0.5 cm need to be identified in a PET scan. Discuss why sophisticated computers are required for PET scanners. *(2 marks)*
6. Discuss the advantages of using Tc-99m as a medical tracer in a patient. *(3 marks)*

27.6 Ultrasound
27.7 Acoustic impedance
27.8 Doppler imaging

Specification reference: 6.5.3

27.6 Ultrasound

Ultrasound is any sound wave with frequency greater than 20 kHz. An **ultrasound transducer** is a device used both to generate and to receive ultrasound.

Ultrasound transducer

Some materials (e.g., quartz) produce an e.m.f. when they are either compressed or extended – this is the **piezoelectric effect**. The effect works in reverse too. When a p.d. is applied across the opposite ends of the material, it will either compress or expand.

The two common piezoelectric materials used in transducers are lead zirconate titanate and polyvinylidene fluoride.

- The piezoelectric material vibrates when a high-frequency alternating p.d. is applied between its opposite ends. The vibration of the material in the air produces ultrasound.
- The same piezoelectric material vibrates when ultrasound is incident on it. These vibrations induce an alternating e.m.f. in the material.

A- and B-scans

In an A-scan, ultrasound pulses from a stationary transducer are sent into the patient. Figure 1 shows the voltage–time trace from the transducer. The voltage pulse 1 is responsible for sending in a pulse of ultrasound into the patient. The ultrasound is partially reflected at the boundaries between the soft tissues. The voltage pulses 2, 3, and 4 are produced from these reflections.

The thickness of tissues A and B can be determined from the average speed c of the ultrasound in A and B and the time t between the voltage pulses using the equation

$$\text{thickness} = \frac{ct}{2}$$

In a B-scan the ultrasound transducer is moved over the patient's skin. The output of the transducer is connected to a computer. For each position of the transducer, the computer produces a row of dots on the digital screen, where each dot corresponds to the boundary between two tissues. The brightness of the dot is proportional to the intensity of the reflected ultrasound pulse. These dots are used to form a two-dimensional image.

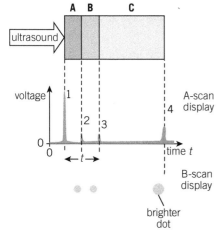

▲ **Figure 1** *Ultrasound A-scan. A, B, and C are different tissues*

> **Revision tip**
> The A in the A-scan stands for 'amplitude' and the B in the B-scan stands for 'brightness'.

27.7 Acoustic impedance

The acoustic impedance Z of a material is defined as the product of its density ρ and the speed c of ultrasound in the material; $Z = \rho c$. The SI unit of acoustic impedance is $\text{kg m}^{-2}\text{s}^{-1}$.

Reflection at a boundary

For a parallel beam of ultrasound incident normally at a boundary between two materials of acoustic impedances Z_1 and Z_2, the ratio of the reflected intensity I_r to the incident intensity I_0 is given by the equation

$$\frac{I_r}{I_0} = \frac{(Z_2 - Z_1)^2}{(Z_2 + Z_1)^2}$$

This ratio is known as the **intensity reflection coefficient**. The acoustic impedance of air is very small compared with that of skin. Placing a transducer directly on the patient's skin would mean most of the ultrasound will be reflected off the skin. To reduce this reflection, a **coupling gel** is smeared on the transducer and the skin. The gel has acoustic impedance similar to that of skin and therefore most of the ultrasound is transmitted into the patient.

The terms impedance matching or acoustic matching are used when two materials have similar acoustic impedances.

27.8 Doppler imaging

The Doppler effect of ultrasound can be used to determine the speed of blood in blood vessels.

Speed of blood

The transducer is placed at an angle θ to the blood vessel, see Figure 2. It sends pulses of ultrasound into the patient. The frequency of the ultrasound reflected off the iron-rich blood cells is changed because of the moving blood cells. The change in frequency Δf is directly proportional to the speed v of the blood. The speed of the blood v can be determined using the equation

$$\Delta f = \frac{2fv\cos\theta}{c}$$

where f is the frequency of the ultrasound from the transducer and c is the speed of ultrasound in blood.

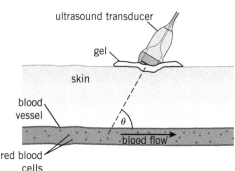

▲ **Figure 2** *An ultrasound transducer can be used to determine the speed of blood*

> **Worked example: Doppler shift**
>
> A transducer emitting ultrasound of frequency 15 MHz is held at an angle of 50° to a blood vessel. The speed of the blood is estimated to be 20 cm s^{-1}. Calculate the change in the frequency of the ultrasound reflected by the blood.
>
> speed of the ultrasound in the blood = 1600 m s^{-1}
>
> **Step 1:** Write down the quantities given.
>
> $f = 15 \times 10^6$ Hz $\quad v = 0.20$ m s^{-1} $\quad c = 1600$ m s^{-1} $\quad \theta = 50°$
>
> **Step 2:** Use the equation $\Delta f = \frac{2fv\cos\theta}{c}$ to calculate Δf.
>
> $$\Delta f = \frac{2fv\cos\theta}{c} = \frac{2 \times 15 \times 10^6 \times 0.20 \times \cos 50°}{1600}$$
>
> $\Delta f = 2.4 \times 10^3$ Hz (2 s.f.)

Summary questions

1 Describe how a transducer emits ultrasound. *(2 marks)*
2 Explain why the distance ct is divided by 2 in the equation thickness $= \frac{ct}{2}$ for an A-scan. *(1 mark)*

3 Suggest why the ultrasound transducer cannot be placed 90° to the blood vessel in Doppler imaging. *(1 mark)*
4 Ultrasound is incident at the boundary between two materials. The acoustic impedance of one of the materials is twice that of the other. Calculate $\frac{I_r}{I_0}$. *(3 marks)*

5 Explain why most of the ultrasound is reflected at the air–skin boundary. *(2 marks)*
6 A transducer emits ultrasound of frequency 18 MHz. It is placed at an angle of 60° to a blood vessel. The change in the frequency of the ultrasound reflected by the blood is 3.2 kHz. The speed of ultrasound in blood is 1600 m s^{-1}. Calculate the speed of the blood in cm s^{-1}. *(3 marks)*

Chapter 27 Practice questions

1 Which technique or device uses the annihilation of electron–positron pairs for imaging?

   A CAT

   B PET

   C A-scan

   D Doppler scan *(1 mark)*

2 The energy of a photon of visible light is about 2 eV.

   What is a good estimate for the energy of an X-ray photon?

   A 10^{-2} eV

   B 10 eV

   C 10^4 eV

   D 10^8 eV *(1 mark)*

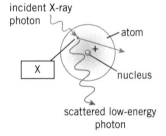

▲ Figure 1

3 Figure 1 shows an incomplete diagram of the Compton effect.

   What is **X** in this figure?

   A electron

   B positron

   C proton

   D vacuum *(1 mark)*

4 a Ultrasound can be used to investigate the internal structures of a patient.

   Explain the A-scan technique. *(4 marks)*

   b Define the acoustic impedance of a material. *(1 mark)*

   c The table below gives some information about two materials X and Y.

Material	Speed of ultrasound in material / m s^{-1}	Density of material / kg m^{-3}	Acoustic impedance /
X	4100	1500	
Y	1600	1100	

   i Complete the table by inserting in the acoustic impedance values for X and Y. Include an appropriate unit. *(3 marks)*

   ii A parallel beam of ultrasound is incident at the boundary of X and Y.

   Calculate the percentage of ultrasound intensity *transmitted* at the boundary. *(3 marks)*

5 a Explain why prolonged exposure to X-rays can be dangerous to a patient. *(2 marks)*

   b A parallel beam of X-rays of intensity 2.0×10^9 W m^{-2} is incident at the soft tissues of a patient for a total time of 1.0 minute.

   Figure 2 shows the variation of $\ln(I/\text{W m}^{-2})$ with thickness x of the tissue, where I is the intensity of the X-rays.

   i Explain why the gradient of the straight-line graph in Figure 2 is $-\mu$, where μ is the attenuation (absorption) coefficient of the soft tissue. *(2 marks)*

   ii Calculate μ in cm^{-1}. *(1 mark)*

   iii Calculate the total energy incident on tissue of cross-sectional area 6.0×10^{-4} m^2 at a depth of 1.8 cm. *(4 marks)*

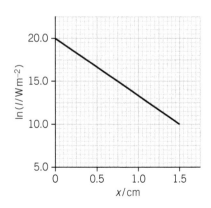

▲ Figure 2

A1 Physical quantities and units

Units

Most of the physical quantities you come across in your course can be expressed by combinations of six base units – kg, m, s, A, K, and mol.

All derived units can be worked out using an appropriate equation and then multiplying and/or dividing the base units.

Example:

density = mass/volume   mass → kg   volume → m^3

Therefore, density has units kg/m^3 or kg m^{-3}.

Homogeneous

An equation is homogeneous when the left-hand side has the same units as the right-hand side. A relationship between physical quantities can only be correct if the equation is homogeneous.

Example:

$s = \frac{1}{2}at^2$

right-hand side: a → m s^{-2}, t^2 → s^2, and $\frac{1}{2}$ has no unit.

Therefore $\frac{1}{2}at^2$ → m s^{-2} × s^2 or m

left-hand side: s → m

The unit m is the same on both sides of the equal sign – the equation is homogeneous.

A2 Recording results and straight lines

Labelling and significant figures

In a table of results, each heading must have the quantity and its unit. The quantity can be represented by a symbol or in words. A slash (solidus /) is used to separate the quantity and its unit, for example, v / m s^{-1} for speed. The labelling of graph axes follows the same rules as for table headings.

Be careful with significant figures in your table of results. The result of a calculation that involves measured quantities has the same number of significant figures as the measurement that has the *smallest* number of significant figures.

Example:

If the distance is 2.12 m (3 s.f.) and time is 3.2 s (2.s.f.), then the speed is written as 0.66 m s^{-1} (2 s.f.).

Graphs

As a general rule, you plot the independent variable, the one you intentionally change in an experiment, on the x-axis, and the dependent variable, the variable which changes as a result, on the y-axis.

A3 Measurements and uncertainties

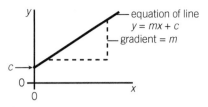

▲ **Figure 1** *Straight-line graph. The gradient of the line is $\frac{\Delta y}{\Delta x}$*

Figure 1 shows a straight-line graph. The equation for a straight line is $y = mx + c$, where m is the gradient and c is the y-intercept. Always use a large triangle to determine the gradient of the straight line.

Sometimes plotting the measured quantities does not produce a straight-line graph. You have to plot data carefully if you want to get a straight line and verify a relationship.

Example:

The equation is $v^2 = 2as$, and you have measured the velocity v and the displacement s. A graph of v against s will be a curve. Plotting v^2 against s should produce a straight line. The gradient of the line will be $2a$, therefore you can determine the acceleration a.

A3 Measurements and uncertainties

Definitions

Remember that no measurement can ever be perfect.

- **Error** (of measurement) is the difference between an individual measurement and the **true** value (or accepted reference value) of the quantity being measured.
- **Random errors** can happen when any measurement is being made. They are measurement errors in which measurements vary unpredictably.
- **Systematic errors** are measurement errors in which the measurements differ from the true values by a consistent amount each time a measurement is made.
- **Accuracy** is to do with how close a measurement result is to the true value.
- **Precision** is to do with how close repeated measurements are to each other.
- The **uncertainty** in the measurement is an interval within which the true value can be expected to lie. The **absolute uncertainty** is approximated as half the range.

Example:

$x = 52 \pm 3\,\text{mm}$

absolute uncertainty = 3 mm

% uncertainty = $\frac{3}{52} \times 100 = 5.8\%$

Uncertainty rules

Rule 1: Adding or subtracting quantities

When you add or subtract quantities in an equation, you add the absolute uncertainties for each value.

Example:

$x = 4.2 \pm 0.3$ $y = 3.0 \pm 0.2$ $x - y = 1.2 \pm 0.5$

Rule 2: Multiplying or dividing quantities

When you multiply or divide quantities, you add the percentage uncertainties for each value.

Example:

$x = 2.0 \pm 0.3 \qquad y = 1.2 \pm 0.1$

% uncertainty in $xy = \left(\dfrac{0.3}{2.0} + \dfrac{0.1}{1.2}\right) \times 100 = 23\%$

Rule 3: Raising a quantity to a power

When a measurement in a calculation is raised to a power n, your percentage uncertainty is increased n times. The power n can be an integer or a fraction.

Example:

$x = 3.14 \pm 0.13$ % uncertainty in $x^4 = 4 \times \left(\dfrac{0.13}{3.14}\right) \times 100 \approx 17\%$

Graphs

You can show error bars for each plotted point, see Figure 2. The best-fit straight line must pass through all the error bars. You can determine the percentage uncertainty in the gradient by drawing a 'worst-fit' straight line and doing the following calculation:

% uncertainty = $\dfrac{\text{gradient of worst-fit line} - \text{gradient of best-fit line}}{\text{gradient of best-fit line}} \times 100\%$

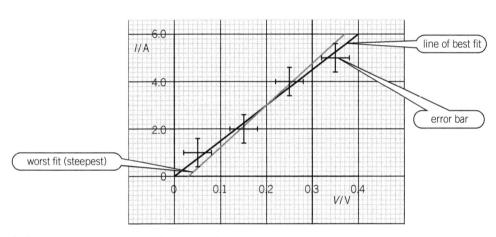

▲ **Figure 2** *Plotted points can have errors bars*

5d. Physics A data sheet

Data, Formulae, and Relationships

The data, formulae, and relationships in this datasheet will be printed for distribution with the examination papers.

Data

Values are given to three significant figures, except where more – or fewer – are useful.

Physical constants

acceleration of free fall	g	9.81 m s^{-2}
elementary charge	e	$1.60 \times 10^{-19} \text{ C}$
speed of light in a vacuum	c	$3.00 \times 10^{8} \text{ m s}^{-1}$
Planck constant	h	$6.63 \times 10^{-34} \text{ J s}$
Avogadro constant	N_A	$6.02 \times 10^{23} \text{ mol}^{-1}$
molar gas constant	R	$8.31 \text{ J mol}^{-1} \text{ K}^{-1}$
Boltzmann constant	k	$1.38 \times 10^{-23} \text{ J K}^{-1}$
gravitational constant	G	$6.67 \times 10^{-11} \text{ N m}^2 \text{ kg}^{-2}$
permittivity of free space	ε_0	$8.85 \times 10^{-12} \text{ C}^2 \text{ N}^{-1} \text{ m}^{-2} \text{ (F m}^{-1})$
electron rest mass	m_e	$9.11 \times 10^{-31} \text{ kg}$
proton rest mass	m_p	$1.673 \times 10^{-27} \text{ kg}$
neutron rest mass	m_n	$1.675 \times 10^{-27} \text{ kg}$
alpha particle rest mass	m_α	$6.646 \times 10^{-27} \text{ kg}$
Stefan constant	σ	$5.67 \times 10^{-8} \text{ W m}^{-2} \text{ K}^{-4}$

Quarks

up quark	charge $= +\frac{2}{3} e$
down quark	charge $= -\frac{1}{3} e$
strange quark	charge $= -\frac{1}{3} e$

Conversion factors

unified atomic mass unit	$1 \text{ u} = 1.661 \times 10^{-27} \text{ kg}$
electronvolt	$1 \text{ eV} = 1.60 \times 10^{-19} \text{ J}$
day	$1 \text{ day} = 8.64 \times 10^{4} \text{ s}$
year	$1 \text{ year} \approx 3.16 \times 10^{7} \text{ s}$
light year	$1 \text{ light year} \approx 9.5 \times 10^{15} \text{ m}$
parsec	$1 \text{ parsec} \approx 3.1 \times 10^{16} \text{ m}$

5d. Physics A data sheet

Mathematical equations

arc length = $r\theta$

circumference of circle = $2\pi r$

area of circle = πr^2

curved surface area of cylinder = $2\pi rh$

surface area of sphere = $4\pi r^2$

area of trapezium = $\frac{1}{2}(a+b)h$

volume of cylinder = $\pi r^2 h$

volume of sphere = $\frac{4}{3}\pi r^3$

Pythagoras' theorem: $a^2 = b^2 + c^2$

cosine rule: $a^2 = b^2 + c^2 - 2bc \cos A$

sine rule: $\frac{a}{\sin A} = \frac{b}{\sin B} = \frac{c}{\sin C}$

$\sin \theta \approx \tan \theta \approx \theta$ and $\cos \theta \approx 1$ for small angles

$\log(AB) = \log(A) + \log(B)$

(Note: $\lg = \log_{10}$ and $\ln = \log_e$)

$\log\left(\frac{A}{B}\right) = \log(A) - \log(B)$

$\log(x^n) = n \log(x)$

$\ln(e^{kx}) = kx$

Formulae and relationships

Module 2 Foundations of physics

vectors

$$F_x = F \cos \theta$$
$$F_y = F \sin \theta$$

Module 3 Forces and motion

uniformly accelerated motion

$$v = u + at$$
$$s = \frac{1}{2}(u+v)t$$
$$s = ut + \frac{1}{2}at^2$$
$$v^2 = u^2 + 2as$$

force

$$F = \frac{\Delta p}{\Delta t}$$
$$p = mv$$

turning effects

$$\text{moment} = Fx$$
$$\text{torque} = Fd$$

density

$$\rho = \frac{m}{V}$$

pressure

$$p = \frac{F}{A}$$
$$p = h\rho g$$

work, energy, and power

$$W = Fx \cos \theta$$
$$\text{efficiency} = \frac{\text{useful energy output}}{\text{total energy input}} \times 100\%$$
$$P = \frac{W}{t}$$
$$P = Fv$$

springs and materials

$$F = kx$$
$$E = \frac{1}{2}Fx; E = \frac{1}{2}kx^2$$
$$\sigma = \frac{F}{A}$$
$$\varepsilon = \frac{x}{L}$$
$$E = \frac{\sigma}{\varepsilon}$$

Module 4 Electrons, waves, and photons

charge

$$\Delta Q = I\Delta t$$

current

$$I = Anev$$

work done

$$W = VQ; W = \mathcal{E}Q; W = VIt$$

resistance and resistors

$$R = \frac{\rho L}{A}$$
$$R = R_1 + R_2 + \ldots$$
$$\frac{1}{R} = \frac{1}{R_1} + \frac{1}{R_2} + \ldots$$

power

$$P = VI; P = I^2R; P = \frac{V^2}{R}$$

internal resistance

$$\mathcal{E} = I(R + r); \mathcal{E} = V + Ir$$

potential divider

$$V_{out} = \frac{R_2}{R_1 + R_2} \times V_{in}$$
$$\frac{V_1}{V_2} = \frac{R_1}{R_2}$$

5d. Physics A data sheet

waves

$$v = f\lambda$$

$$f = \frac{1}{T}$$

$$I = \frac{P}{A}$$

$$\lambda = \frac{ax}{D}$$

refraction

$$n = \frac{c}{v}$$

$$n \sin \theta = \text{constant}$$

$$\sin C = \frac{1}{n}$$

quantum physics

$$E = hf$$

$$E = \frac{hc}{\lambda}$$

$$hf = \phi + KE_{max}$$

$$\lambda = \frac{h}{p}$$

Answers to practice questions

Chapter 14

1 C [1]   2 B [1]   3 A [1]

4 a $E = mc\Delta\theta = 0.120 \times 4200 \times 15$ [1]
     rate of energy loss $= \dfrac{0.120 \times 4200 \times 15}{6 \times 60}$ [1]
     rate of energy loss $= 21\,W$ [1]

  b $E = 3.3 \times 10^5 \times 0.120$ [1]
     time $= \dfrac{3.3 \times 10^5 \times 0.120}{21}$ [1]
     time $= 1900\,s$ (31 mins) [1]

5 a   A graph of m against t plotted with correct labelling. [1]
     A line of best fit is drawn through the data points. [1]

  b   gradient $= 9.3 \times 10^{-5}\,kg\,s^{-1}$
     (Allow $\pm 0.1 \times 10^{-5}\,kg\,s^{-1}$) [1]

  c $Pt = mL_f$ therefore gradient of m–t graph is $\dfrac{P}{L_f}$. [1]
     $\dfrac{25}{L_f} = 9.3 \times 10^{-5}$ [1]
     $L_f = 2.7 \times 10^5\,J\,kg^{-1}$ [1]

  d   The melting ice is also absorbing energy from the surroundings. [1]

6 a   The temperature of the block is greater than that of the water, so there is a net transfer of thermal energy from the block to the water. [1]

  b $E = mc\Delta\theta = 0.500 \times 4200 \times 17$ [1]
     $E = 3.57 \times 10^4\,J \approx 3.6 \times 10^4\,J$ [1]

  c $3.57 \times 10^4 = 0.210 \times 450 \times (\theta - 37)$ [1]
     $\theta - 37 = 377.8$ [1]
     $\theta = 414.8 \approx 410\,°C$ [1]

Chapter 15

1 B [1]   2 D [1]   3 C [1]   4 A [1]

5 a $pV = nRT$, with $n = 1$ therefore $pV = RT$ [1]

  b $p \propto T$ therefore a graph of p against T will be a straight line (through the origin). [1]

  c   gradient $= \dfrac{R}{V} = \dfrac{N_A k}{V}$ [1]
     gradient $= 375\,Pa\,K^{-1}$ (Allow $\pm 15\,Pa\,K^{-1}$) [1]
     $k = \dfrac{375 \times 2.2 \times 10^{-2}}{6.02 \times 10^{23}} = 1.37 \times 10^{-23}\,J\,K^{-1} \approx 1.4 \times 10^{-23}\,J\,K^{-1}$ [1]

6 a   density $= \dfrac{M}{V} = \dfrac{0.65 \times 10^{-3}}{5.2 \times 10^{-4}} = 1.25\,kg\,m^{-3}$ [1]

  b $n = \dfrac{0.65 \times 10^{-3}}{30 \times 10^{-3}} = 2.167 \times 10^{-2}\,mol \approx 2.2 \times 10^{-2}\,mol$ [1]

  c $PV = nRT$ [1]
     $T = \dfrac{1.2 \times 10^5 \times 5.2 \times 10^{-4}}{8.31 \times 2.167 \times 10^{-2}} = 346.5\,K$ [1]
     temperature $= 346.5 - 273 \approx 74\,°C$ [1]

  d   mean KE $= \dfrac{3}{2}kT = \dfrac{3}{2} \times 1.38 \times 10^{-23} \times 346.5$ [1]
     mean KE $= 7.2 \times 10^{-21}\,J$ [1]

Chapter 16

1 D [1]   2 D [1]   3 B [1]   4 D [1]

5 a   Its velocity is changing. [1]
     Acceleration is the rate of change of velocity – the object must therefore have acceleration. [1]

  b   i   The arrow on the object is towards the centre of the disc. [1]
     ii   weight $= mg = 0.120 \times 9.81$ [1]
       friction $= F = 0.50 \times 0.120 \times 9.81$ [1]
       $F = \dfrac{mv^2}{r}; v = \sqrt{\dfrac{Fr}{m}} = \sqrt{\dfrac{0.50 \times 0.120 \times 9.81 \times 0.10}{0.120}}$ [1]
       $v = 0.70\,m\,s^{-1}$ [1]

6 a $a = \dfrac{v^2}{r} = \dfrac{(4.2 \times 10^7)^2}{0.012}$ [1]
     $a = 1.47 \times 10^{17}\,m\,s^{-2} \approx 1.5 \times 10^{17}\,m\,s^{-2}$ [1]

  b $F = ma = 9.11 \times 10^{-31} \times 1.47 \times 10^{17}$ [1]
     $F = 1.34 \times 10^{-13}\,N \approx 1.3 \times 10^{-13}\,N$ [1]

  c $T = \dfrac{2\pi r}{v} = \dfrac{2\pi \times 0.012}{4.2 \times 10^7}$ [1]
     $T = 1.8 \times 10^{-9}\,s$ [1]

7 a $T\cos 45° = 1.2; T = 1.697\,N \approx 1.7\,N$ [1]

  b   centripetal force $= T\sin 45° = 1.697 \times \sin 45° = 1.2\,N$ [1]

  c $F = \dfrac{mv^2}{r}$; mass $m = \dfrac{1.2}{9.81} = 0.1223\,kg$ [1]
     $v = \sqrt{\dfrac{Fr}{m}} = \sqrt{\dfrac{1.2 \times 0.15}{0.1223}}$ [1]
     $v = 1.2\,m\,s^{-1}$ [1]

Chapter 17

1 D [1]   2 D [1]   3 B [1]   4 B [1]

5 a $a \propto -x$ [1]

  b $A = 1.6\,mm$ [1]

  c   i $v_{max} = \omega A = 2\pi f A = 2\pi \times 840 \times 1.6 \times 10^{-3}$ [1]
       $v_{max} = 8.446\,m\,s^{-1} \approx 8.4\,m\,s^{-1}$ [1]
     ii $a = \omega^2 A = (2\pi \times 840)^2 \times 1.6 \times 10^{-3}$ [1]
       $a = 4.5 \times 10^4\,m\,s^{-2}$ [1]

  d   The potential energy is zero at the equilibrium position. [1]
     It increases as the displacement increases, reaching a maximum at the amplitude position. [1]

6 a $v_{max} = \omega A$ [1]

  b $v_{max} = \omega A$, where ω is constant and therefore v_{max} is proportional to A.
     This gives a straight line (through the origin). [1]

Answers to practice questions

    c i gradient = $8.0\,s^{-1}$ (Allow $\pm 0.1\,s^{-1}$) [1]

       ii gradient = $\omega = 2\pi f$ [1]

          $f = \dfrac{8.0}{2\pi}$ [1]

          $f = 1.27\,\text{Hz} \approx 1.3\,\text{Hz}$ [1]

Chapter 18

1 C [1]   **2** A [1]   **3** C [1]

4   a   gravitational field strength $g = \dfrac{\text{force}}{\text{mass}} = \dfrac{F}{m}$

       and acceleration $a = \dfrac{\text{force}}{\text{mass}} = \dfrac{F}{m}$ [1]

       Therefore $a = g$. [1]

    b   i g is inversely proportional to r^2 therefore $gr^2 =$ constant. [1]

          A minimum of <u>two</u> from:

          $9.8 \times (6400 \times 10^3)^2 = 4.0 \times 10^{14}$

          $6.3 \times (8000 \times 10^3)^2 = 4.0 \times 10^{14}$

          $4.4 \times (9600 \times 10^3)^2 = 4.0 \times 10^{14}$ [1]

       ii $gr^2 = GM = 4.0 \times 10^{14}$ [1]

          $M = \dfrac{4.0 \times 10^{14}}{6.67 \times 10^{-11}}$ [1]

          $M = 6.0 \times 10^{24}\,\text{kg}$ [1]

       iii   density $= \dfrac{M}{V} = \dfrac{6.0 \times 10^{24}}{\frac{4}{3}\pi \times (6400 \times 10^3)^3}$ [1]

          density $= 5.5 \times 10^3\,\text{kg}\,\text{m}^{-3}$ [1]

5   a $\dfrac{GMm}{r^2} = \dfrac{mv^2}{r}$ [2]

       $v = \sqrt{\dfrac{GM}{r}}$ [1]

    b   ratio $= \sqrt{\dfrac{r_{\text{Neptune}}}{r_{\text{Earth}}}}$ [1]

       ratio $= \sqrt{30} = 5.5$ [1]

6   a   2.8 MJ [1]

    b $V_g = -\dfrac{GM}{r} \propto \dfrac{1}{r}$ [1]

       V_g at $3R = -\dfrac{2.8}{3} = -0.933\,\text{MJ}\,\text{kg}^{-1} \approx -9.3 \times 10^5\,\text{J}\,\text{kg}^{-1}$ [1]

       V_g at $4R = -\dfrac{2.8}{4} = -0.70\,\text{MJ}\,\text{kg}^{-1} \approx -7.0 \times 10^5\,\text{J}\,\text{kg}^{-1}$ [1]

    c   energy $= \Delta V_g \times m = (0.933 - 0.70) \times 10^6 \times 3500$ [1]

       energy $= 8.2 \times 10^8\,\text{J}$ [1]

Chapter 19

1 D [1]   **2** A [1]   **3** D [1]

4   a   Any <u>two</u> from:

       Hot surface temperature, remnant of a low-mass star, mass less than 1.44 solar masses, no fusion takes place, etc. [2]

    b   i   Correct shape of intensity against wavelength curve, see below. [1]

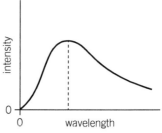

       ii   wavelength at maximum intensity $\times$ temperature in K = constant (Wien's displacement law) [1]

    c $L = 4\pi r^2 \sigma T^4$ [1]

       $L = 3.8 \times 10^{26} \times \dfrac{1}{120^2} \times \left(\dfrac{25000}{5800}\right)^4$ [1]

       $L = 9.1 \times 10^{24}\,\text{W}$ [1]

5   a   The emission (or absorption) spectrum of each element is unique. [1]

       By measuring the wavelengths using a grating, these elements can be identified in the stars. [1]

    b $d\sin\theta = n\lambda$; $\theta = 90°$ and $d = \dfrac{10^{-3}}{800} = 1.25 \times 10^{-6}\,\text{m}$ [1]

       $n = \dfrac{1.25 \times 10^{-6}}{656 \times 10^{-9}} = 1.905$ [1]

       Therefore, maximum order n is 1. [1]

    c $\sin\theta = \dfrac{1 \times 589 \times 10^{-9}}{1.25 \times 10^{-6}}$; $\theta = 28.11°$ [1]

       $\sin\theta = \dfrac{1 \times 587 \times 10^{-9}}{1.25 \times 10^{-6}}$; $\theta = 28.01°$ [1]

       angular separation $= 0.10°$ [1]

6   a   Luminosity is the total radiant power emitted from the surface of a star. [1]

    b   i $L = 4\pi r^2 \sigma T^4 = 4\pi \times (8.5 \times 10^8)^2 \times 5.67 \times 10^{-8} \times (5800)^4$ [1]

          $L = 5.83 \times 10^{26}\,\text{W}$ [1]

       ii   intensity $= \dfrac{5.83 \times 10^{26}}{4\pi \times (3.9 \times 10^{16})^2}$ [1]

          intensity $= 3.05 \times 10^{-8}\,\text{W}\,\text{m}^{-2}$ [1]

    c   A sensible suggestion, e.g., absorption of radiation by dust or Earth's atmosphere. [1]

Chapter 20

1 C [1]   **2** D [1]   **3** D [1]   **4** C [1]

5   a   Correct diagram, see below. [1]

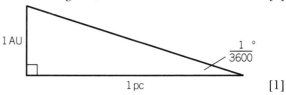

[1]

       $\tan\left(\dfrac{1}{3600}°\right) = \dfrac{1.5 \times 10^{11}}{1\,\text{pc}}$ [1]

Answers to practice questions

$1 \text{ pc} = \dfrac{1.5 \times 10^{11}}{\tan(2.78 \times 10^{-4})°} = 3.09 \times 10^{16} \text{ m} \approx$
$3.1 \times 10^{16} \text{ m}$ [1]

b distance $= 4.0 \times 10^4 \times 3.1 \times 10^{16} = 1.24 \times 10^{21} \text{ m}$ [1]

time $= \dfrac{1.24 \times 10^{21}}{3.0 \times 10^8} = 4.13 \times 10^{12} \text{ s} \approx 4.1 \times 10^{12} \text{ s}$ [1]

time $= \dfrac{4.13 \times 10^{12}}{3.16 \times 10^7} = 1.3 \times 10^5 \text{ y}$ [1]

c i Hubble constant $= \dfrac{70 \times 10^3}{3.1 \times 10^{16} \times 10^6}$ [1]

Hubble constant $H_0 = 2.26 \times 10^{-18} \text{ s}^{-1} \approx$
$2.3 \times 10^{-18} \text{ s}^{-1}$ [1]

ii age $= H_0^{-1} = (2.26 \times 10^{-18})^{-1} = 4.43 \times 10^{17} \text{ s}$ [1]

The maximum distance is the product of the speed of light and the maximum time. [1]

distance $= 4.43 \times 10^{17} \times 3.0 \times 10^8 =$
$1.33 \times 10^{26} \text{ m} \approx 1.3 \times 10^{26} \text{ m}$ [1]

d Any two from:
Microwave background radiation,
3K temperature, and abundance of helium. [2]

6 a Line of best fit drawn (must pass through all the error bars). [1]

b $\dfrac{\Delta \lambda}{\lambda} \approx \dfrac{v}{c}$, where c and v are constants therefore

$\Delta \lambda \propto \lambda$ and therefore a straight line (through the origin). [1]

c gradient $= 6.8 \times 10^{-5}$ (Allow $\pm 0.2 \times 10^{-5}$) [1]

$\dfrac{v}{c} = \dfrac{v}{3.00 \times 10^8} = 6.8 \times 10^{-5}$ [1]

$v = 2.0 \times 10^4 \text{ m s}^{-1}$ [1]

Chapter 21

1 B [1]   **2** D [1]   **3** A [1]   **4** D [1]

5 a Any four from:

The current in the circuit charges the capacitor.

The charge stored by the capacitor increases and therefore the p.d. across it increases.

The sum of the p.d.s across the capacitor and resistor must add to 6.00 V.

Therefore, as the p.d. across the capacitor increases, the p.d. across the resistor decreases.

The final p.d. across the capacitor is 6.00 V and zero across the resistor. [4]

b $V_c = V_0(1 - e^{-\tfrac{t}{CR}})$ and
$CR = 120 \times 10^{-6} \times 1.0 \times 10^6 = 120 \text{ s}$ [1]

$V_c = 6.00(1 - e^{-\tfrac{200}{120}})$ [1]

$V_c = 4.87 \text{ V}$ [1]

c maximum current $= \dfrac{6.00}{1.0 \times 10^6} = 6.00 \mu\text{A}$ [1]

d $E = \dfrac{1}{2}V^2C = \dfrac{1}{2} \times 6.00^2 \times 120 \times 10^{-6}$ [1]

$E = 2.2 \times 10^{-3} \text{ J}$ [1]

6 a $CR = 150 \times 10^{-6} \times 100 \times 10^3 = 15 \text{ s}$ [1]

b $V = V_0 e^{-\tfrac{t}{CR}} = 4.50 \times e^{-\tfrac{35}{15}}$ [1]

$V = 0.44 \text{ V}$ [1]

c energy dissipated = difference in the energy stored by the capacitor [1]

energy dissipated $= \dfrac{1}{2}C(V_1^2 - V_2^2) =$

$\dfrac{1}{2} \times 150 \times 10^{-6}(4.50^2 - 0.44^2)$ [1]

energy dissipated $= 1.5 \times 10^{-3} \text{ J}$ [1]

Chapter 22

1 C [1]   **2** A [1]   **3** C [1]   **4** C [1]

5 a $E \propto \dfrac{1}{r^2}$; therefore Er^2 = constant [1]

A minimum of two from:

$20.7 \times 10^3 \times 0.092^2 = 175$

$5.4 \times 10^3 \times 0.18^2 = 175$

$2.8 \times 10^3 \times 0.25^2 = 175$ [1]

b $E = \dfrac{Q}{4\pi\varepsilon_0 r^2}$ therefore $Er^2 = \dfrac{Q}{4\pi\varepsilon_0}$ [1]

$Q = 4\pi \times 8.85 \times 10^{-12} \times 175$ [1]

$Q = 1.946 \times 10^{-8} \text{ C} \approx 1.9 \times 10^{-8} \text{ C}$ [1]

c charge per unit area $= \dfrac{Q}{4\pi r^2}$ [1]

charge per init area $= \dfrac{1.946 \times 10^{-8}}{4\pi \times 0.050^2} =$
$6.2 \times 10^{-7} \text{ C m}^{-2}$ [1]

d $V = \dfrac{Q}{4\pi\varepsilon_0 r} = \dfrac{1.946 \times 10^{-8}}{4\pi \times 8.85 \times 10^{-12} \times 0.05}$ [1]

$V = 3.5 \times 10^3 \text{ V}$ [1]

6 a The electric field is uniform between the plates, with field lines parallel and equally spaced (and perpendicular to the plates). [1]

The electric field strength is constant between the plates. [1]

b $E = \dfrac{V}{d} = \dfrac{5000}{0.072}$ [1]

$E = 6.94 \times 10^4 \text{ V m}^{-1} \approx 6.9 \times 10^4 \text{ V m}^{-1}$ [1]

c $a = \dfrac{EQ}{m} = \dfrac{6.94 \times 10^4 \times 1.60 \times 10^{-19}}{9.11 \times 10^{-31}} =$
$1.22 \times 10^{16} \text{ m s}^{-2}$ [2]

$v^2 = u^2 + 2as = 0 + 2 \times 1.22 \times 10^{16} \times 0.072$ [1]

$v = 4.2 \times 10^7 \text{ m s}^{-1}$ [1]

d The final kinetic energy of the electron is Ve. [1]

This is independent of the separation between the plates. [1]

Answers to practice questions

Chapter 23

1 A [1] 2 B [1] 3 B [1]

4 a $f = T^{-1} = 0.020^{-1} = 50\,\text{Hz}$ [1]

  b $F = BIL$; $4.0 \times 10^{-3} = 0.060 \times I \times 0.01$ [1]

  $I = 6.7\,\text{A}$ [1]

  c $F = BIL\sin\theta$ therefore F will decrease as θ is decreased. [1]

  The graph will remain sine-shaped, but maximum force will decrease. [1]

5 a $BQv = \dfrac{mv^2}{r}$ [1]

  $mv = p = BQr$ [1]

  b i $p = Ber = 0.012 \times 1.60 \times 10^{-19} \times 0.032$ [1]

  $p = 6.144 \times 10^{-23}\,\text{kg m s}^{-1} \approx 6.1 \times 10^{-23}\,\text{kg m s}^{-1}$ [1]

  ii $v = \dfrac{6.144 \times 10^{-23}}{9.11 \times 10^{-31}} = 6.744 \times 10^{7}\,\text{m s}^{-1}$ [1]

  $\text{KE} = \dfrac{1}{2} \times 9.11 \times 10^{-31} \times (6.744 \times 10^{7})^2 = 2.07 \times 10^{-15}\,\text{J}$ [1]

  $\text{KE} = 13\,\text{keV}$ [1]

6 a When the current is switched on, an increasing magnetic flux links the coil. [1]

  The induced e.m.f. is equal to the rate of change of magnetic flux therefore an e.m.f. is induced in the coil for a short period. [1]

  When the current is constant, the flux linkage is constant and therefore no e.m.f. is induced in the coil. [1]

  b Field pattern with concentric circles and separation between field lines increasing as shown. [2]

current into plane of paper

Chapter 24

1 A [1] 2 B [1] 3 A [1] 4 C [1]

5 a R is proportional to $A^{\frac{1}{3}}$, therefore $\dfrac{R}{\sqrt[3]{A}} = $ constant. [1]

  A minimum of <u>two</u> data used, e.g., $\dfrac{5.6}{\sqrt[3]{100}} = 1.2$ and $\dfrac{7.0}{\sqrt[3]{200}} = 1.2$ [1]

  b The mass is proportional to A. [1]

  The volume is $\dfrac{4}{3}\pi r^3$ therefore volume is proportional to A. [1]

  The density (which is mass divided by volume) is independent of A. [1]

  c mass $= 4.00 \times 1.66 \times 10^{-27}\,\text{kg}$ [1]

  $r = 1.2 \times 4^{\frac{1}{3}} = 1.9\,\text{fm} = 1.9 \times 10^{-15}\,\text{m}$

  [The 1.2 fm is from part (a)] [1]

  density $= \dfrac{4.00 \times 1.66 \times 10^{-27}}{\dfrac{4}{3}\pi \times (1.9 \times 10^{-15})^3}$ [1]

  density $= 2.3 \times 10^{17}\,\text{kg m}^{-3}$ [1]

6 a The repulsive electrostatic force on the alpha-particle is greater at shorter distances, and therefore has greater deviation in its path. [1]

  b electrical potential energy = kinetic energy = $8.2 \times 10^{-13}\,\text{J}$ [1]

  $\dfrac{Qq}{4\pi\varepsilon_0 r} = \dfrac{79 \times 2 \times (1.60 \times 10^{-19})^2}{4\pi \times 8.85 \times 10^{-12} \times r} = 8.2 \times 10^{-13}$ [1]

  $r = 4.44 \times 10^{-14}\,\text{m} \approx 4.4 \times 10^{-14}\,\text{m}$ [1]

  c $F = \dfrac{Qq}{4\pi\varepsilon_0 r^2} = \dfrac{79 \times 2 \times (1.60 \times 10^{-19})^2}{4\pi \times 8.85 \times 10^{-12} \times (4.4 \times 10^{-14})^2}$ [1]

  force $= 19\,\text{N}$ [1]

Chapter 25

1 B [1] 2 B [1] 3 C [1] 4 B [1]

5 a Exponential decay has a constant-ratio property. [1]

  With $\Delta t = 5.0\,\text{s}$, $\dfrac{34}{40} \approx \dfrac{28}{34} \approx \dfrac{24}{28}$ [1]

  b $A_0 = 40 \times 10^{14}\,\text{Bq}$ and $A = 24 \times 10^{14}\,\text{Bq}$ when $t = 15\,\text{s}$ [1]

  $A = A_0 e^{-\lambda t}$; $24 = 40 e^{-\lambda \times 15}$ [1]

  $\lambda = -\dfrac{\ln(0.60)}{15} = 3.41 \times 10^{-2}\,\text{s}^{-1}$ [1]

  $T = \dfrac{\ln 2}{3.41 \times 10^{-2}} = 20.3\,\text{s} \approx 20\,\text{s}$ [1]

  c $^{10}_{6}\text{C} \rightarrow {}^{10}_{5}\text{B} + {}^{0}_{+1}\text{e} + \nu_e$ (Allow $^{10}_{5}\text{X}$ for $^{10}_{5}\text{B}$) [2]

  d power = activity 3.4×10^{-13} [1]

  power $= 40 \times 10^{14} \times 3.4 \times 10^{-13} = 1.36 \times 10^{3}\,\text{W} \approx 1.3\,\text{kW}$ [1]

6 a The count rate is a fraction of the total activity of the source. [1]

  Therefore, the suggestion is incorrect. [1]

  b i $\lambda = \dfrac{\ln 2}{T} = \dfrac{\ln 2}{6.0 \times 3600} = 3.21 \times 10^{-5}\,\text{s}^{-1}$ [1]

  $A = \lambda N = 3.21 \times 10^{-5} \times 2.0 \times 10^{9}$ [1]

  $A = 6.4 \times 10^{4}\,\text{Bq}$ [1]

  ii 12 hours = 2 half-lives [1]

  $N = \dfrac{2.0 \times 10^{9}}{4} = 5.0 \times 10^{8}$ [1]

  iii $N = N_0 e^{-\lambda t} = 2.0 \times 10^{9} \times e^{-(3.21 \times 10^{-5} \times 21 \times 3600)}$ [1]

  $N = 1.8 \times 10^{8}$ [1]

Chapter 26

1 A [1] 2 D [1] 3 B [1] 4 C [1]

5 a This isotope has the largest BE per nucleon and therefore it cannot decay. [1]

Answers to practice questions

b i The BE per nucleon of the $_1^2$H nucleus is less than the BE per nucleon of the $_2^4$He nucleus. [1]

The BE of the $_2^4$He nucleus is greater than the total BE of the $_1^2$H nuclei. [1]

The difference in the BE is released as energy. [1]

ii number of $_1^2$H 'pairs' = 3.01×10^{23} [1]

difference in BE ≈ $4 \times 7.1 - 4 \times 1.0 = 24.4$ MeV [1]

total energy released = $3.01 \times 10^{23} \times 24.4 \times 10^6 \times 1.60 \times 10^{-19}$ [1]

total energy = 1.2×10^{12} J [1]

6 a The change in energy ΔE is related to the change in mass Δm by the equation $\Delta E = \Delta mc^2$, where c is the speed of light in a vacuum. Therefore, mass and energy are equivalent. [1]

b i The hydrogen nuclei are positively charged and therefore repel each other. [1]

High temperatures are necessary for the nuclei to have high enough KE to get close enough to each other so that the strong nuclear force can fuse together the hydrogen nuclei. [1]

ii The mass of the helium-3 nucleus is less than the total mass of the hydrogen-2 and hydrogen-1 nuclei. [1]

The difference in the mass is equivalent to the energy released ($\Delta E = \Delta mc^2$). [1]

iii energy = $5.5 \times 10^6 \times 1.60 \times 10^{-19}$ [1]

$\Delta m = \dfrac{5.5 \times 10^6 \times 1.60 \times 10^{-19}}{(3.00 \times 10^8)^2}$ [1]

$\Delta m = 9.8 \times 10^{-30}$ kg [1]

c In fission, a neutron is absorbed by a 'heavy' nucleus (e.g., uranium-235). This causes the nucleus to *split* into smaller nuclei and fast moving neutrons. [1]

Fusion occurs with 'lighter' nuclei (e.g., hydrogen, helium, etc.). These lighter nuclei *join together* to form a larger nucleus. [1]

Chapter 27

1 B [1]   **2** C [1]   **3** A [1]

4 a A transducer is used to send a pulse of ultrasound into the patient. [1]

The pulse is reflected at the boundaries between tissues. [1]

The reflected pulse is monitored using the transducer. [1]

The time difference t between the pulse received and sent is used to find the depth x of the boundary; $x = \dfrac{1}{2}ct$ where c is the (average) speed of sound in the tissues. [1]

4 b acoustic impedance = density of material × speed of ultrasound in material [1]

c i $Z_X = 1500 \times 4100 = 6.15 \times 10^6$ [1]

$Z_Y = 1100 \times 1600 = 1.76 \times 10^6$ [1]

unit: kg m^{-2} s^{-1} [1]

ii $\dfrac{I_r}{I_0} = \dfrac{(Z_2 - Z_1)^2}{(Z_2 + Z_1)^2} = \dfrac{(6.15 - 1.76)^2}{(6.15 + 1.76)^2}$ [1]

$\dfrac{I_r}{I_0} = 0.31$ [1]

percentage transmitted = 69% [1]

5 a X-rays cause ionisation. [1]

It can damage healthy cells in the patient. [1]

b i $I = I_0 e^{-\mu x}$ therefore $\ln I = \ln I_0 - \mu x$ [1]

Compare with the equation for a straight line, $y = mx + c$. The gradient is $-\mu$. [1]

ii gradient = 6.7 cm^{-1} (Allow ±0.1 cm^{-1}) [1]

iii $I = I_0 e^{-\mu x} = 2.0 \times 10^9 e^{-6.7 \times 1.8}$ [1]

$I = 1.157... \times 10^4$ W m^{-2} [1]

power = $1.157... \times 10^4 \times 6.0 \times 10^{-4} = 6.944$ W [1]

energy = $1.0 \times 60 \times 6.944 = 420$ J [1]

Answers to summary questions

14.1/14.2

1. Any sensible comment, e.g., temperature is measured in °C or K and heat energy is measured in J. [1]

2. Insert the thermometer into pure *melting* ice. The level of mercury in the glass will be the 0°C mark. [1]

3. a 3 K [1]
   b 221 K [1]
   c 293 K [1]
   d 373 K [1]
   e 2273 K [1]

4. a −270 °C [1]
   b 0 °C [1]
   c 107 °C [1]
   d 227 °C [1]
   e 5227 °C [1]

5. The potential energy of the molecules is greater in the water phase because of the increased separation between the molecules. [2]

   The kinetic energy of the molecules is also greater in the water phase because of the increase in the temperature. [2]

6. The mean kinetic energy of the molecules and the smoke particles is the same at a specific temperature. [1]

   $KE = \frac{1}{2}mv^2$ therefore $v \propto \frac{1}{\sqrt{m}}$ [1]

   The mass m of the molecules << mass of the smoke particles therefore mean speed of molecules >> mean speed of the smoke particles. [1]

14.3/14.4/14.5

1. Absolute zero is temperature of 0 K. The internal energy of a substance at absolute zero is a minimum. [1]

2. $E = mL_f = 0.010 \times 3.3 \times 10^4$ [1]
   $E = 330$ J [1]

3. $E = mc\Delta\theta = 0.300 \times 4200 \times (100 - 20) = 1.01 \times 10^5$ J [1]
   time $= \frac{\text{energy}}{\text{power}} = \frac{1.01 \times 10^5}{200}$ [1]
   time = 504 s (8.4 minutes) [1]

4. $E = mL_v = 0.300 \times 2.3 \times 10^6 = 6.9 \times 10^5$ J [1]
   time $= \frac{\text{energy}}{\text{power}} = \frac{6.9 \times 10^5}{200}$ [1]
   time = 3.45×10^3 s (58 minutes) [1]

5. energy lost by water = $E = mc\Delta\theta = 0.120 \times 4200 \times (100 - 87) = 6552$ J [1]
   $6552 = 0.200 \times c \times (87 - 20)$ [1]
   $c = \frac{6552}{13.4}$ [1]
   $c = 490$ J kg⁻¹ K⁻¹ [1]

6. energy needed to change phase = $0.500 \times 3.3 \times 10^4 = 1.65 \times 10^4$ J [1]
   energy needed to warm water = $0.500 \times 4200 \times 20 = 4.2 \times 10^4$ J [1]
   time $= \frac{\text{energy}}{\text{power}} = \frac{1.65 \times 10^4 + 4.2 \times 10^4}{120}$ [1]
   time = 490 s (8.1 minutes) [1]

15.1/15.2

1. number of molecules = $3.0 \times 6.02 \times 10^{23} = 1.8 \times 10^{24}$ [1]

2. mass of molecule = $\frac{0.032}{6.02} \times 10^{23}$ [1]
   mass of molecule = 5.3×10^{-26} kg [1]

3. $p \propto T$ [1]
   T increases by a factor of $\frac{473}{373} = 1.27 \approx 1.3$ [1]
   Therefore, the pressure p also increases by a factor of 1.3. [1]

4. Assuming the temperature of the water is constant, volume $\propto \frac{1}{\text{pressure}}$ [1]
   As the bubble rises, the pressure exerted on it by the water decreases ($p = h\rho g$). [1]
   Therefore, the volume of the bubble increases as it rises. [1]

5. $pV = nRT$ [1]
   $V = \frac{nRT}{p} = \frac{1 \times 8.31 \times [273 - 50]}{1.0 \times 10^5}$ [1]
   $V = 1.853 \times 10^{-2}$ m³ [1]
   density $= \frac{0.029}{1.853} \times 10^{-2} = 1.6$ kg m⁻³ [1]

6. Any reasonable estimates: $V = 25$ m³, $T = 293$ K (20 °C), and $p = 1.0 \times 10^5$ Pa. [1]
   $n = \frac{pV}{RT} = \frac{1.0 \times 10^5 \times 25}{8.31 \times 293} = 1.027 \times 10^3$ mol [2]
   mass = $0.029 \times 1.027 \times 10^3 \approx 30$ kg [1]

15.3/15.4

1. mean velocity $= \frac{-100 - 200 + 150 + 200 + 300}{5}$
   $= +70$ m s⁻¹ [1]
   mean speed $= \frac{100 + 200 + 150 + 200 + 300}{5}$
   $= 190$ m s⁻¹ [1]

2. mean square speed
   $= \frac{100^2 + 200^2 + 150^2 + 200^2 + 300^2}{5} = 40\,500$ m² s⁻² [1]
   r.m.s. speed = $\sqrt{40500} = 200$ m s⁻¹ [1]

3. mean KE of molecules
   $= \frac{3}{2}kT = \frac{3}{2} \times 1.38 \times 10^{-23} \times (273 + 200)$ [1]
   mean KE of molecules = 9.8×10^{-21} J [1]

Answers to summary questions

4 The smoke particles and the air molecules have the same mean kinetic energy at a specific temperature. [1]

As the mass of the smoke particle >> mass of the air molecule, according to $\overline{c^2} = \dfrac{3kT}{m}$, the r.m.s. speed of the 'heavier' smoke particles << r.m.s. of the air molecules. [1]

5 $\dfrac{1}{2}m\overline{c^2} = \dfrac{3}{2}kT = 9.8 \times 10^{-21}$ [1]

$\dfrac{1}{2} \times 4.8 \times 10^{-26} \times \overline{c^2} = 9.8 \times 10^{-21}$ [1]

r.m.s. speed = $640\,\text{m s}^{-1}$ [1]

6 internal energy = $2.0 \times 6.02 \times 10^{23} \times \dfrac{3}{2}kT$ [1]

internal energy
= $2.0 \times 6.02 \times 10^{23} \times \dfrac{3}{2} \times 1.38 \times 10^{-23} \times 273$ [1]

internal energy = $6.8 \times 10^{3}\,\text{J}$ [1]

16.1/16.2

1 **a** 0.79 rad [1] **b** 0.087 rad [1] **c** 7.3 rad [1]

2 $v = \omega r = 15 \times 0.50$ [1]
   $v = 7.5\,\text{m s}^{-1}$ [1]

3 $45° = 0.785$ rad and $\omega = \dfrac{\theta}{t}$ [1]
   $\omega = \dfrac{0.785}{2.5} = 0.31\,\text{rad s}^{-1}$ [1]

4 $v = \omega r;\ 150 = \omega \times 12 \times 10^{3}$ [1]
   $\omega = 0.0125\,\text{rad s}^{-1} \approx 0.013\,\text{rad s}^{-1}$ [1]
   $a = \dfrac{v^2}{r} = \dfrac{150^2}{12 \times 10^{3}}$ [1]
   $a = 1.9\,\text{m s}^{-2}$ [1]

5 $a = g = \dfrac{v^2}{r};\ 9.81 = \dfrac{20^2}{r}$ [1]
   $r = 41\,\text{m}$ [1]

6 The centripetal force is at right angles to the velocity. [1]

The centripetal force has no component in the direction of the velocity ($F\cos 90 = 0$) – no work is done on the object and therefore the speed does not change. [1]

16.3

1 Gravitational force between the planet and Sun. [1]

2 $F = \dfrac{mv^2}{r} = \dfrac{0.20 \times 3.2^2}{0.50}$ [1]
   $F = 4.1\,\text{N}$ [1]

3 $4300 = \dfrac{mv^2}{r} = \dfrac{900 \times v^2}{30}$ [1]
   $v = 12\,\text{m s}^{-1}$ [1]

4 $\dfrac{mv^2}{r} = Mg$ or $v^2 = \dfrac{gr}{m} \times M$ [1]

The equation for a straight line through the origin is $y = mx$ therefore the gradient is equal to $\dfrac{gr}{m}$. [1]

5 $v = \dfrac{2\pi r}{T} = \dfrac{2\pi \times 3.8 \times 10^{8}}{27 \times 24 \times 3600} = 1024\,\text{m s}^{-1}$ [1]
   $F = \dfrac{mv^2}{r} = \dfrac{7.3 \times 10^{22} \times 1024^2}{3.8 \times 10^{8}}$ [1]
   $F = 2.0 \times 10^{20}\,\text{N}$ [1]

6 $L\sin\theta = \dfrac{mv^2}{r}$ and $L\cos\theta = W = mg$ [1]

Divide these two equations to give $\dfrac{\sin\theta}{\cos\theta} = \dfrac{v^2}{rg}$ [1]

$\tan\theta = \dfrac{\sin\theta}{\cos\theta}$ therefore $\tan\theta = \dfrac{v^2}{rg}$ [1]

17.1

1 $\omega = \dfrac{2\pi}{T} = \dfrac{2\pi}{2.0} = 3.14\,\text{rad s}^{-1}$ [1]

2 At the equilibrium position ($x = 0$), the acceleration is zero and it increases to a maximum when it is maximum displacement. [1]

The direction of the acceleration is always towards the equilibrium position. [1]

3 $\omega = 2\pi f = 2\pi \times 1500$ [1]
   $\omega = 9400\,\text{rad s}^{-1}$ [1]
   $a_{\max} = \omega^2 A = (2\pi \times 1500)^2 \times 0.60 \times 10^{-3}$ [1]
   $a_{\max} = 5.3 \times 10^{4}\,\text{m s}^{-2}$ [1]

4 $\phi = 2\pi \times \left(\dfrac{\Delta t}{T}\right) = 2\pi \times \dfrac{1}{8}$ [1]
   $\phi = 0.79\,\text{rad}$

5 $\omega^2 = 400$ [1]
   $\dfrac{2\pi}{T} = \sqrt{400}$ [1]
   $T = 0.31\,\text{s}$ [1]

6 $\omega^2 = \dfrac{g}{L}$ [1]
   $a_{\max} = \dfrac{g}{L} \times A$ [1]
   $F_{\max} = \dfrac{mgA}{L} = \dfrac{1.2 \times 9.81 \times 1.6}{5.0}$ [1]
   $F_{\max} = 3.8\,\text{N}$ [1]

17.2/17.3

1 The maximum speed is directly proportional to the amplitude ($v_{\max} = \omega A$). [1]

2 The acceleration against time graph is an 'inverted' displacement–time graph. [1]

Therefore, acceleration ∝ –displacement, illustrating SHM. [1]

3 $v_{\max} = \omega A = \dfrac{2\pi}{0.040} \times 3.0 \times 10^{-3}$ [1]
   $v_{\max} = 0.47\,\text{m s}^{-1}$ [1]

4 $v = \omega\sqrt{A^2 - x^2} = \dfrac{2\pi}{0.040} \times (0.003^2 - 0.001^2)^{\tfrac{1}{2}}$ [1]
   $v = 0.44\,\text{m s}^{-1}$ [1]

5 $E_k = \dfrac{1}{2}m\omega^2(A^2 - x^2)$ therefore $E_p = \dfrac{1}{2}m\omega^2 x^2$ [1]
   $E_p = \dfrac{1}{2} \times 0.180 \times \left(\dfrac{2\pi}{0.80}\right)^2 \times 0.10^2$ [1]
   $E_p = 0.056\,\text{J}$ [1]

Answers to summary questions

6 $x = A\cos\omega t$; $0.075 = 0.200 \times \cos\dfrac{2p}{0.80 \times t}$ [1]

$7.86t = \cos^{-1}(0.375)$

$t = 0.15\,\text{s}$ [1]

17.4/17.5

1 An oscillator is in resonance when it reaches maximum amplitude when being forced to oscillate. [1]

2 The amplitude of the oscillations decreases with time (because of frictional forces). [1]

3 The sharpness of the resonance decreases with increased damping. [1]

4 The resonant frequency becomes less than the natural frequency as the amount of damping is increased. [1]

5 $\dfrac{1}{8} = \dfrac{1}{2^3}$ therefore number of oscillations = 10×3 [1]

number of oscillations = 30 [1]

6 total energy = $\dfrac{1}{2}m\omega^2 A^2 \propto A^2$ [1]

total energy = $10^2 \times 0.0126$ [1]

total energy = $1.26\,\text{J} \approx 1.3\,\text{J}$ [1]

18.1/18.2/18.3

1 $g = \dfrac{F}{m} = \dfrac{190}{50} = 3.8\,\text{N kg}^{-1}$ [1]

2 $F = mg = 600 \times 3.0 = 1.8 \times 10^3\,\text{N}$ [1]

3 $F = \dfrac{GMm}{r^2} = \dfrac{6.67 \times 10^{-11} \times 2.0 \times 10^{30} \times 3.3 \times 10^{23}}{(5.8 \times 10^{10})^2}$ [1]

$F = 1.3 \times 10^{22}\,\text{N}$ [1]

4 $F \propto \dfrac{1}{r^2}$ [1]

The distance r has increased by a factor of **4**. [1]

force = $\dfrac{F}{4^2} = \dfrac{F}{16}$ [1]

5 $g = \dfrac{GM}{r^2}$; $r = \sqrt{\dfrac{GM}{g}}$ [1]

$r = \sqrt{\dfrac{6.67 \times 10^{-11} \times 8.7 \times 10^{25}}{10}}$ [1]

radius = $2.4 \times 10^7\,\text{m}$ [1]

6 $g \propto \dfrac{M}{r^2}$ [1]

$g = 9.81 \times \dfrac{320}{11^2}$ [1]

$g = 26\,\text{N kg}^{-1}$ [1]

18.4/18.5

1 The square of the orbital period of any planet is directly proportional to the cube of the mean distance from the Sun. [1]

2 The orbital period is 1 day. [1]

3 $\dfrac{GMm}{r^2} = \dfrac{mv^2}{r}$ [1]

$v^2 = \dfrac{GM}{r} = \dfrac{6.67 \times 10^{-11} \times 6.0 \times 10^{24}}{(6400 \times 10^3 + 3000 \times 10^3)}$ [2]

$v = 6.5 \times 10^3\,\text{m s}^{-1}$ [1]

4 period = $\dfrac{\text{circumference}}{\text{speed}}$ [1]

period = $\dfrac{2\pi \times (6400 + 3000) \times 10^3}{6.5 \times 10^3}$ [1]

period = $9100\,\text{s}$ (2.5 hours) [1]

5 Kepler's third law: $T^2 = kr^3$ [1]

Take logs of both sides: $\lg T^2 = \lg k + \lg r^3$

$2\lg T = \lg k + 3\lg r$

$\lg T = \dfrac{\lg k}{2} + 1.5\lg r$ [1]

The equation for a straight line is $y = mx + c$, therefore a graph of $\lg T$ against $\lg r$ will be a straight line of gradient 1.5. [1]

6 $T^2 = \left(\dfrac{4\pi^2}{GM}\right)r^3$ [1]

$(24 \times 3600)^2 = \left(\dfrac{4\pi^2}{6.67 \times 10^{-11} \times 6.0 \times 10^{24}}\right) \times r^3$

$r = 4.23 \times 10^7\,\text{m}$ [1]

$r = \dfrac{4.23 \times 10^7}{6400 \times 10^3} = 6.6$.

The radius is 6.6 times the Earth radius. [1]

18.6/18.7

1 Zero. [1]

2 54 MJ [1]

3 $V_g = -\dfrac{GM}{r}$; $54 \times 10^6 = \dfrac{6.67 \times 10^{-11} \times M}{6.1 \times 10^6}$ [1]

$M = 4.9 \times 10^{24}\,\text{kg}$ [1]

4 $V_g = -\dfrac{GM}{r}$; $V_g = -\dfrac{6.67 \times 10^{-11} \times 6.0 \times 10^{24}}{6400 \times 10^3}$ [1]

$V_g = -6.3 \times 10^7\,\text{J kg}^{-1}$ [1]

5 $\Delta V_g =$

$-6.67 \times 10^{-11} \times 6.0 \times 10^{24}\left(\dfrac{1}{9400 \times 10^3} - \dfrac{1}{6400 \times 10^3}\right)$ [2]

$\Delta V_g = 2.0 \times 10^7\,\text{J kg}^{-1}$ [1]

6 change in GPE = $2.0 \times 10^7 \times 1200$ [1]

change in GPE = $2.4 \times 10^{10}\,\text{J}$ [1]

19.1/19.2/19.3

1 $1.44 \times$ solar mass [1]

2 $10 \times 2.0 \times 10^{30} = 2.0 \times 10^{31}\,\text{kg}$ [1]

3 The mass of the core. (The remnant is a neutron star if the mass of the core $< 3M_\odot$ and a black hole if $> 3M_\odot$.) [1]

4 Red giants have lower temperature and larger luminosity than white dwarfs. [2]

Answers to summary questions

5 density = $\frac{mass}{volume} = \frac{4.0 \times 10^{30}}{\frac{4}{3}\pi \times 11000^3} = 7.2 \times 10^{17}\,kg\,m^{-3}$ [1]

mass = density × volume = $7.2 \times 10^{17} \times 10^{-9}$ [1]

mass = $7.2 \times 10^8\,kg$ [1]

6 $v = \sqrt{\frac{2GM}{r}}$, $v = c$ [1]

$r = \frac{2GM}{c^2} = \frac{2 \times 6.67 \times 10^{-11} \times 5 \times 2.0 \times 10^{30}}{(3.0 \times 10^8)^2}$ [1]

radius = 15 km [1]

19.4/19.5

1 An energy level is one of the discrete set of energies of an electron inside an atom. [1]

2 energy of photon = 6.0 − 2.0 = 4.0 eV [1]

3 The energy of the electron increases therefore a photon is absorbed by the electron. [1]

4 energy of photon = $4.0 \times 1.6 \times 10^{-19} = 6.4 \times 10^{-19}\,J$ [1]

$\frac{hc}{\lambda} = \Delta E$; $\lambda = \frac{6.63 \times 10^{-34} \times 3.0 \times 10^8}{6.4 \times 10^{-19}}$ [1]

$\lambda = 3.1 \times 10^{-7}\,m$ [1]

5 There are 6 possible spectral lines. [1]

See energy level diagram below.

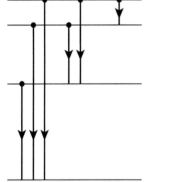

[1]

6 $\frac{hc}{\lambda} = \Delta E$; $\Delta E = \frac{6.63 \times 10^{-34} \times 3.0 \times 10^8}{400 \times 10^{-9}} = 4.97 \times 10^{-19}\,J$ [2]

$\Delta E = \frac{4.97 \times 10^{-19}}{1.60 \times 10^{-19}} = 3.11\,eV$ [1]

energy level = −12.2 + 3.11

energy level = −9.1 eV [1]

19.6/19.7

1 $d = \frac{10^{-3}}{80} = 1.25 \times 10^{-5}\,m$ [1]

2 luminosity ∝ (thermodynamic temperature)^4 [1]

3 $d = \frac{10^{-3}}{800} = 1.25 \times 10^{-6}\,m$ [1]

$d\sin\theta = n\lambda$; $\sin\theta = \frac{1 \times 6.4 \times 10^{-7}}{1.25 \times 10^{-6}}$ [1]

$\theta = 30.8° \approx 31°$ [1]

4 $\lambda_{max}T$ = constant; $500 \times 5800 = \lambda_{max} \times 12000$ [1]

$\lambda_{max} = 240\,nm$ [1]

5 $L = 4\pi r^2 \sigma T^4 \propto r^2$ [1]

Red giants have greater radius (and therefore surface area). Therefore, the total power emitted is much greater than the smaller main sequence stars. [1]

6 For the Sun: $L_\odot = 4\pi R_\odot^2 \sigma \times 5800^4$ [1]

For Rigel: $L = 4\pi(79\,R_\odot)^2 \sigma \times 12000^4$ [1]

$L = \frac{79^2 \times 12000^4}{5800^4} L_\odot$ [1]

$L = 1.14 \times 10^5 L_\odot$ [1]

20.1/20.2

1 a distance = $5.2 \times 1.5 \times 10^{11} = 7.8 \times 10^{11}\,m$ [1]

b distance = $1.6 \times 9.5 \times 10^{15} = 1.52 \times 10^{16}\,m \approx 1.5 \times 10^{16}\,m$ [1]

c distance = $2600 \times 9.5 \times 10^{15} = 2.47 \times 10^{19}\,m \approx 2.5 \times 10^{19}\,m$ [1]

d distance = $95 \times 3.1 \times 10^{16} = 2.945 \times 10^{18}\,m \approx 2.9 \times 10^{18}\,m$ [1]

2 angle = $\frac{0.78}{3600} = 2.2 \times 10^{-4\,\circ}$ [1]

3 distance = $8.0 \times 10^3 \times 3.1 \times 10^{16} = 2.48 \times 10^{20}\,m$ [1]

time = $\frac{2.48 \times 10^{20}}{3.0 \times 10^8}$ [1]

time = $8.3 \times 10^{11}\,s\ (2.6 \times 10^4\,y)$ [1]

4 $\frac{\Delta\lambda}{\lambda} \approx \frac{v}{c} = \frac{7.6 \times 10^3}{3.0 \times 10^8} = 2.5 \times 10^{-5}$ [1]

% change = 2.5×10^{-3} % [1]

5 distance in pc = $8.6 \times \frac{9.5 \times 10^{15}}{3.1 \times 10^{16}}$ [1]

distance = 2.64 pc [1]

$p = \frac{1}{d} = \frac{1}{2.64} = 0.38$ arcseconds [1]

6 $\frac{\Delta\lambda}{\lambda} \approx \frac{v}{c} = \frac{5.3 \times 10^6}{3.0 \times 10^8} = 1.767 \times 10^{-2}$ [1]

$\Delta\lambda = 1.767 \times 10^{-2} \times 119.5 = 2.1\,nm$ [1]

Observed wavelength = 119.5 + 2.1 = 121.6 nm [1]

20.3/20.4/20.5

1 2.7 K [1]

2 When viewed on a large enough scale, the Universe is homogeneous and isotropic, and the laws of physics are universal. [1]

3 $v = H_0 d = 70 \times 200$ [1]

$v = 14000\,km\,s^{-1}$ [1]

4 $5200 = 70 \times d$ [1]

$d = 74\,Mpc$ [1]

5 $70\,km\,s^{-1}\,Mpc^{-1} = \frac{70 \times 10^3}{10^6 \times 3.1 \times 10^{16}}$ [2]

$70\,km\,s^{-1}\,Mpc^{-1} = 2.26 \times 10^{-18}\,s^{-1}$ [1]

6 $\lambda_{max}T$ = constant; $T = 5800\,K$ and $\lambda_{max} = 5.0 \times 10^{-7}\,m$ for the Sun [1]

$\lambda_{max} = \frac{5.0 \times 10^{-7} \times 5800}{2.7} = 1.1 \times 10^{-3}\,m$ [1]

This wavelength is in the microwave region of the EM spectrum. [1]

Answers to summary questions

21.1/21.2/21.3

1. $Q = VC = 9.0 \times 500 \times 10^{-6}$ [1]
   $Q = 4.5 \times 10^{-3}$ C [1]

2. energy $= \frac{1}{2}V^2C = \frac{1}{2} \times 9.0^2 \times 500 \times 10^{-6}$ [1]
   energy $= 2.0 \times 10^{-2}$ J [1]

3. maximum: parallel combination
   with $C = 100 + 100 + 100 = 300\,\mu\text{F}$ [2]
   minimum: series combination with $C = \frac{100}{3} = 33\,\mu\text{F}$ [2]

4. parallel branch: $C = 300 + 100 = 400\,\mu\text{F}$ [1]
   complete circuit: $C = (400^{-1} + 500^{-1})^{-1} = 220\,\mu\text{F} \approx 220\,\mu\text{F}$ [2]

5. charge on the $500\,\mu\text{F}$ capacitor:
   $Q = 6.0 \times 222 \times 10^{-6} = 1.33 \times 10^{-3}$ C [1]
   p.d. $= \frac{Q}{C} = \frac{1.33 \times 10^{-3}}{500} \times 10^{-6}$ [1]
   p.d. $= 2.7$ V [1]

6. charge $= 10 \times 1000 \times 10^{-6} = 1.0 \times 10^{-2}$ C [1]
   total capacitance $= C = 1000 + 500 = 1500\,\mu\text{F}$ [1]
   The p.d. across each capacitor is the same. [1]
   $V = \frac{Q}{C} = \frac{1.0 \times 10^{-2}}{1500 \times 10^{-6}} = 6.7$ V [1]

21.4/21.5/21.6

1. time constant $= CR = \tau$
   $\tau = 100 \times 10^{-6} \times 150 \times 10^3 = 15$ s [1]

2. There is a current in the circuit and charge is removed from the capacitor. [1]
   Since charge $\propto$ p.d., the p.d. across the capacitor decreases. [1]

3. $CR = 100 \times 10^{-6} \times 200 \times 10^3 = 20$ s [1]
   $V = V_0 e^{-\frac{t}{CR}} = 10 \times e^{-\frac{38}{20}}$ [1]
   $V = 1.5$ V [1]

4. $t = 5CR$ [1]
   $Q = Q_0 e^{-\frac{t}{CR}} = Q_0 \times e^{-5}$ [1]
   $Q = 0.0067\,Q_0$ therefore charge left is 0.67% < 1%. [1]

5. $CR = 500 \times 10^{-6} \times 100 \times 10^6 = 50$ s [1]
   $V_C = V_0(1 - e^{-\frac{t}{CR}}) = 10 \times (1 - e^{-\frac{80}{50}}) = 8.0$ V [1]
   $V_R = 10.0 - 8.0$ [1]
   $V_R = 2.0$ V [1]

6. $V = V_0 e^{-\frac{t}{CR}}$; $CR = 20$ s and $V = 0.5\,V_0$ [1]
   $0.5 = e^{-\frac{t}{20}}$ [1]
   $\ln 0.5 = -\frac{t}{20}$ [1]
   $t = 14$ s [1]

22.1/22.2

1. $E = \frac{F}{Q} = \frac{8.0 \times 10^{-14}}{1.60 \times 10^{-19}}$ [1]
   $E = 5.0 \times 10^5\,\text{N C}^{-1}$ [1]

2. $F = EQ = 6.0 \times 10^4 \times 1.60 \times 10^{-19}$ [1]
   $F = 9.6 \times 10^{-15}$ N [1]

3. $F = \frac{Qq}{4\pi\varepsilon_0 r^2} = \frac{(1.60 \times 10^{-19})^2}{4\pi \times 8.85 \times 10^{-12} \times (2.0 \times 10^{-10})^2}$ [2]
   $F = 5.76 \times 10^{-9}\,\text{N} \approx 5.8 \times 10^{-9}$ N [1]

4. $F \propto \frac{1}{r^2}$ therefore the force will increase by a factor of 4. [1]
   force $= 4 \times 5.76 \times 10^{-9} = 2.3 \times 10^{-8}$ N [1]

5. $E = \frac{Q}{4\pi\varepsilon_0 r^2}$
   $r = \sqrt{\frac{Q}{4\pi\varepsilon_0 E}} = \sqrt{\frac{3.8 \times 10^{-9}}{4\pi \times 8.85 \times 10^{-12} \times 5.0 \times 10^4}}$ [2]
   $r = 0.026$ m (2.6 cm) [1]

6. $E = \frac{Q}{4\pi\varepsilon_0 r^2} = \left(\frac{Q}{4\pi r^2}\right) \times \frac{1}{\varepsilon_0}$
   (surface area of sphere $= 4\pi r^2$) [1]
   Therefore, $E = \frac{\sigma}{\varepsilon_0}$ [1]
   $\sigma = \varepsilon_0 E$ [1]

22.3/22.4

1. $E = \frac{V}{d} = \frac{2000}{0.01}$ [1]
   $E = 2.0 \times 10^5\,\text{V m}^{-1}$ [1]

2. $Q = VC = 2000 \times 10 \times 10^{-12} = 2.0 \times 10^{-8}$ C [1]

3. $C = \frac{\varepsilon_0 A}{d}$; $10 \times 10^{-12} = \frac{8.85 \times 10^{-12} \times A}{0.01}$ [1]
   $A = 1.13 \times 10^{-2}\,\text{m}^2 \approx 1.1 \times 10^{-2}\,\text{m}^2$ [1]

4. $C = \frac{\varepsilon_r \varepsilon_0 A}{d} = \frac{4.0 \times 8.85 \times 10^{-12} \times 6.24 \times 10^{-2}}{0.070 \times 10^{-3}}$
   $= 31.6$ nF [1]
   $C = 3.16 \times 10^{-8}$ F [1]
   $C = 32$ nF [1]

5. $F = EQ = \frac{500}{0.01} \times 1.60 \times 10^{-19} = 8.0 \times 10^{-15}$ N [1]
   $a = \frac{8.0 \times 10^{-15}}{9.11 \times 10^{-31}} = 8.782 \times 10^{15}\,\text{m s}^{-2}$ [1]
   $s = \frac{1}{2}at^2$; $0.01 = \frac{1}{2} \times 8.782 \times 10^{15} \times t^2$ [1]
   $t = 1.5 \times 10^{-9}$ s [1]

6. $E = \frac{1}{2}\frac{Q^2}{C}$ and $C = \varepsilon_0 \frac{A}{d}$ [1]
   $E \propto d$ (Q and A are constants) [1]
   As the separation is doubled, the final energy stored is $2E_0$. [1]

Answers to summary questions

22.5

1. energy = 100 J [1]
2. $V = \dfrac{Q}{4\pi\varepsilon_0 r} \propto \dfrac{1}{r}$ [1]

   The distance from the centre is doubled therefore $V = \dfrac{2000}{2} = 1000\,\text{V}$ [1]

3. $V = \dfrac{Q}{4\pi\varepsilon_0 r} = \dfrac{1.60 \times 10^{-19}}{4\pi \times 8.85 \times 10^{-12} \times 1.2 \times 10^{-10}}$ [1]

   $V = 12\,\text{V}$ [1]

4. $V = \dfrac{Q}{4\pi\varepsilon_0 r}$; $Q = V \times 4\pi\varepsilon_0 r$ [1]

   $Q = 100 \times 4\pi \times 8.85 \times 10^{-12} \times 0.050$ [1]

   $Q = 5.6 \times 10^{-10}\,\text{C}$ [1]

5. $C = 4\pi\varepsilon_0 R = 4\pi \times 8.85 \times 10^{-12} \times 0.025$ [1]

   $C = 2.78 \times 10^{-12}\,\text{F}$ [1]

   $Q = VC = 5000 \times 2.78 \times 10^{-12}$ [1]

   $Q = 1.4 \times 10^{-8}\,\text{C}$ [1]

6. $Q = V \times 4\pi\varepsilon_0 r$ [1]

   $Q = 3000 \times 4\pi \times 8.85 \times 10^{-12} \times 0.015 = 5.00 \times 10^{-9}\,\text{C}$ [1]

   $F = \dfrac{Qq}{4\pi\varepsilon_0 r^2} = \dfrac{(5.00 \times 10^{-9})^2}{4\pi \times 8.85 \times 10^{-12} \times 0.050^2}$ [1]

   $F = 9.0 \times 10^{-5}\,\text{N}$ [1]

23.1/23.2

1. The magnetic flux density is constant. [1]
2. The magnetic field is similar to that of a bar magnetic. [1]

   However, within the core, the magnetic field is uniform (parallel and equally spaced magnetic field lines). [1]

3. $F = BIL = 0.020 \times 5.0 \times 0.040$ [1]

   $F = 4.0 \times 10^{-3}\,\text{N}$ [1]

4. $F = BIL \propto BI$ [1]

   Therefore, the force on the wire will be $\dfrac{3}{2}F$. [1]

5. Each wire lies in the magnetic field of the other wire. [1]

   Therefore, each wire will experience a magnetic force.

6. $\sin\theta = \dfrac{F}{BIL} = \dfrac{1.5 \times 10^{-3}}{0.02 \times 5.0 \times 0.040}$ [1]

   $\theta = 22°$ [1]

23.3

1. The force is perpendicular to the velocity of the electron. [1]

   Therefore, no work is done on the electron – its speed remains constant. [1]

2. $F = BQv = 0.080 \times 1.60 \times 10^{-19} \times 4.0 \times 10^6$ [1]

   $F = 5.12 \times 10^{-14}\,\text{N} \approx 5.1 \times 10^{-14}\,\text{N}$ [1]

3. $a = \dfrac{F}{m} = \dfrac{5.12 \times 10^{-14}}{9.11 \times 10^{-31}}$ [1]

   $a = 5.62 \times 10^{16}\,\text{m s}^{-2} \approx 5.6 \times 10^{16}\,\text{m s}^{-2}$ [1]

4. $a = \dfrac{v^2}{r}$; $5.62 \times 10^{16} = \dfrac{(4.0 \times 10^6)^2}{r}$ [1]

   $r = 2.8 \times 10^{-4}\,\text{m}$ [1]

5. $r = \dfrac{mv}{BQ} = \dfrac{6.7 \times 10^{-27} \times 6.0 \times 10^5}{0.720 \times 2 \times 1.60 \times 10^{-19}}$ [2]

   $r = 0.017\,\text{m}$ [1]

6. $T = \dfrac{2\pi m}{Be}$ (see worked example) [1]

   $T = \dfrac{2\pi \times 1.673 \times 10^{-27}}{0.900 \times 1.60 \times 10^{-19}}$ [1]

   $T = 7.3 \times 10^{-8}\,\text{s}$ [1]

23.4/23.5/23.6

1. $1\,\text{Wb} = 1\,\text{T m}^2$ [1]
2. The secondary (output) coil has more turns than the primary (input) coil. [1]
3. The magnetic flux linking the coil is constant. [1]

   There is no induced e.m.f. because the rate of change of magnetic flux linkage is zero. [1]

4. $\dfrac{n_s}{n_p} = \dfrac{V_s}{V_p} = \dfrac{5}{230} = \dfrac{1}{46}$ [1]

   The primary coil has 46 times more turns than the secondary coil. [1]

5. The rate of change of magnetic flux increases. [1]

   The maximum e.m.f. would increase. [1]

   The frequency of the output signal would also increase. [1]

6. $\varepsilon = \dfrac{\Delta(N\phi)}{\Delta t} = \dfrac{\Delta(BAN)}{\Delta t}$ (magnitude only) [1]

   $0.010 = \dfrac{B \times 2.0 \times 10^{-4} \times 500}{0.20}$ [1]

   $B = 0.020\,\text{T}$ [1]

24.1/24.2

1. There are 8 protons and 7 neutrons. [2]
2. The nucleus is 10^5 times smaller than the atom. [1]

   The chance of an alpha particle getting close to the nucleus, and therefore be deflected, is very small. [1]

3. Helium nucleus: $R = r_0 A^{\frac{1}{3}} = 1.2 \times 4^{\frac{1}{3}} = 1.9\,\text{fm}$ [1]

   Uranium nucleus: $R = r_0 A^{\frac{1}{3}} = 1.2 \times 235^{\frac{1}{3}} = 7.4\,\text{fm}$ [1]

4. The radius of an atom is about $10^{-10}\,\text{m}$ or diameter of about $2 \times 10^{-10}\,\text{m}$. [1]

   If connected end-to-end:

   number of atoms $= \dfrac{0.010}{2 \times 10^{-10}} = 5 \times 10^7$ atoms [1]

5. At the minimum separation, the alpha particle stops momentarily therefore initial kinetic energy = final electric potential energy. [1]

   kinetic energy = 7.7 MeV = $7.7 \times 10^6 \times 1.60 \times 10^{-19} = 1.232 \times 10^{-12}\,\text{J}$ [1]

Answers to summary questions

$E = \dfrac{Qq}{4\pi\varepsilon_0 r}$ or $r = \dfrac{Qq}{4\pi\varepsilon_0 E}$ [1]

$r = \dfrac{13 \times 2 \times (1.60 \times 10^{-19})^2}{4\pi \times 8.85 \times 10^{-12} \times 1.232 \times 10^{-12}} = 4.86 \times 10^{-15}\,\text{m} \approx 4.9 \times 10^{-15}\,\text{m}$ [1]

6 $F = \dfrac{Qq}{4\pi\varepsilon_0 r^2} = \dfrac{13 \times 2 \times (1.60 \times 10^{-19})^2}{4\pi \times 8.85 \times 10^{-12} \times (4.86 \times 10^{-15})^2}$ [2]

$F = 250\,\text{N}$ [1]

24.3/24.4/24.5

1 A fundamental particle has no internal structure and cannot be sub-divided into smaller particles. [1]
Any two correct examples, e.g., quark and electron. [1]

2 Quarks experience the strong nuclear force. [1]

3 $^{1}_{0}\text{n} \rightarrow {^{1}_{1}}\text{p} + {^{0}_{-1}}\text{e} + \overline{\nu}_e$ [2]
Quantities conserved: nucleon number and proton number (or charge). [1]

4 The strong nuclear force experienced by the protons is short-ranged (~10^{-15}m) and attractive. [1]
The gravitational force on the protons has an infinite range and is attractive. [1]

5 neutron → u d d [1]
charge $= \left(\dfrac{2}{3} - \dfrac{1}{3} - \dfrac{1}{3}\right)e = 0$ [1]

6 u u d → u d d + $^{0}_{+1}$e + ν_e [2]

7 Antimatter coming into contact with matter will annihilate each other. [1]
Since we observe matter around us and in the Universe, there must be more matter than antimatter. [1]

8 $F_G = \dfrac{GMm}{r^2}$ and $F_E = \dfrac{Qq}{4\pi\varepsilon_0 r^2}$ [1]

ratio $= \dfrac{4\pi\varepsilon_0 GMm}{Qq}$ [1]

ratio $= \dfrac{4\pi \times 8.85 \times 10^{-12} \times 6.67 \times 10^{-11} \times (1.7 \times 10^{-27})^2}{(1.6 \times 10^{-19})^2}$ [1]

ratio $\approx 8 \times 10^{-37}$ [1]

25.1/25.2

1 $^{0}_{-1}$e is an electron. [1]
$\overline{\nu}_e$ is an electron antineutrino. [1]

2 a The two numbers conserved are proton number and nucleon number. [1]
b number of neutrons in magnesium-28 nucleus = 28 − 12 = 16 [1]
number of neutrons in aluminium-28 nucleus = 28 − 13 = 15 [1]

3 a $^{204}_{82}\text{Pb} \rightarrow {^{200}_{80}}\text{Hg} + {^{4}_{2}}\text{He}$ [1]
b $^{249}_{98}\text{Cf} \rightarrow {^{245}_{96}}\text{Cm} + {^{4}_{2}}\text{He}$ [1]

4 a $^{19}_{8}\text{O} \rightarrow {^{19}_{9}}\text{F} + {^{0}_{-1}}\text{e} + \overline{\nu}_e$ [1]
b $^{21}_{11}\text{Na} \rightarrow {^{21}_{10}}\text{Ne} + {^{0}_{+1}}\text{e} + \nu_e$ [2]

5 The number of gamma photons emitted per second from the source is spread equally over a sphere of radius 30 cm and the GM tube detects a fraction of this count rate. [1]
number of photons per second $= \dfrac{4\pi \times 0.30^2}{2.0 \times 10^{-4}} \times 120$ [1]
number of photons per second $= 6.8 \times 10^5\,\text{s}^{-1}$ [1]

6 $10\,\text{eV} = 10 \times 1.6 \times 10^{-19}\,\text{J} = 1.6 \times 10^{-18}\,\text{J}$ [1]
kinetic energy of alpha particle $= 1.6 \times 10^{-18} \times 10^4 \times 30 = 4.8 \times 10^{-13}\,\text{J}$ [1]
$\dfrac{1}{2} \times 6.6 \times 10^{-27} \times v^2 = 4.8 \times 10^{-13}$ [1]
$v = 1.2 \times 10^7\,\text{m s}^{-1}$ [1]

25.3/25.4

1 The decay constant is inversely proportional to half-life. [1]

2 number of alpha particles $= 120 \times 2 = 240$ [1]
Assumption: The activity remains constant over the period of 2.0 s. [1]

3 a 4.0 mins is 2 half-lives. [1]
number of nuclei left $= \dfrac{8.0 \times 10^{15}}{2^2} = 2.0 \times 10^{15}$ [1]
b 6.0 mins is 8 half-lives. [1]
number of nuclei left $= \dfrac{8.0 \times 10^{15}}{2^3} = 1.0 \times 10^{15}$ [1]
number of nuclei decayed $= (8.0 - 1.0) \times 10^{15} = 7.0 \times 10^{15}$ [1]

4 $1.5\,\text{MeV} = 1.5 \times 10^6 \times 1.60 \times 10^{-19}\,\text{J}$ [1]
power = activity × energy of each α-particle [1]
power $= 3.4 \times 10^{10} \times 1.5 \times 10^6 \times 1.60 \times 10^{-19} = 8.16 \times 10^{-3}\,\text{W} \approx 8.2\,\text{mW}$ [1]

5 number of nuclei $= \dfrac{1.5 \times 10^{-6}}{0.234} \times 6.02 \times 10^{23} = 3.86 \times 10^{18}$ [1]
$\lambda = \dfrac{\ln(2)}{6.7 \times 3600} = 2.87 \times 10^{-5}\,\text{s}^{-1}$ [1]
$A = \lambda N = 2.87 \times 10^{-5} \times 3.86 \times 10^{18}$ [1]
$A = 1.1 \times 10^{14}\,\text{Bq}$ [1]

6 $\lambda = \dfrac{\ln(2)}{5730} = 1.21 \times 10^{-4}\,\text{y}^{-1}$ [1]
$A = A_0 e^{-\lambda t}$; $0.72 = e^{-1.21 \times 10^{-4} t}$ [1]
$\ln(0.72) = -1.21 \times 10^{-4} t$ [1]
$t = \dfrac{\ln(0.72)}{-1.21 \times 10^{-4}} = 2720\,\text{y}$ [1]

25.5/25.6

1 The activity of the source is not constant over such a long period of time. [1]

2 a $\lambda = \dfrac{\ln(2)}{10} = 0.0693\,\text{s}^{-1} \approx 0.069\,\text{s}^{-1}$ [1]
b number of nuclei decaying $= \Delta N = (\lambda \Delta t) N$ [1]
$\Delta N = 0.0693 \times 0.10 \times 1000 = 6.9$ (allow 7) [1]

Answers to summary questions

3 The activity has decreased by a factor of 4 – this is equivalent to 2 half-lives. [1]
   age = $5730 \times 2 \approx 11500$ y [1]

4 $\lambda = \dfrac{\ln(2)}{5730} = 1.21 \times 10^{-4}$ y^{-1} [1]
   $A = \lambda N$; $N = \dfrac{1.7}{1.21 \times 10^{-4}}$ [1]
   $N = 1.4 \times 10^4$ [1]

5 $\lambda = \dfrac{\ln(2)}{5730} = 1.21 \times 10^{-4}$ y^{-1} [1]
   $A = A_0 e^{-\lambda t}$; $0.32 = 1.6 e^{-1.21 \times 10^{-4} t}$ [1]
   $t = -\dfrac{\ln(0.20)}{1.21 \times 10^{-4}}$ [1]
   $t = 13300$ y ≈ 13000 y [1]

6 $\lambda = \dfrac{\ln(2)}{49 \times 10^9} = 1.41 \times 10^{-11}$ y^{-1} [1]
   $\dfrac{N}{N_0} = e^{-\lambda t}$; $0.94 = e^{-1.41 \times 10^{-11} t}$ [1]
   $t = -\dfrac{\ln(0.94)}{1.41 \times 10^{-11}}$ [1]
   $t = 4.4 \times 10^9$ y   (4.4 billion years) [1]

26.1/26.2

1 $\Delta E = \Delta m c^2 = 1.0 \times 10^{-6} \times (3.00 \times 10^8)^2$ [1]
   energy $= 9.0 \times 10^{10}$ J [1]

2 $\Delta E = \Delta m c^2 = 9.11 \times 10^{-31} \times (3.0 \times 10^8)^2$ [1]
   energy $= 8.20 \times 10^{-14}$ J [1]

3 a The (thermal) energy of the lump of iron decreases. [1]
      Therefore, its mass will *decrease* because $\Delta E \propto \Delta m$. [1]
   b The (kinetic) energy of the electron decreases. [1]
      Therefore, its mass will *decrease* because $\Delta E \propto \Delta m$. [1]
   c The (kinetic) energy of the proton increases. [1]
      Therefore, its mass will *increase* because $\Delta E \propto \Delta m$. [1]

4 a BE per nucleon = 1.0 MeV [1]
      BE = $1.0 \times 2 = 2.0$ MeV [1]
   b BE per nucleon = 7.1 MeV [1]
      BE = $7.1 \times 4 = 28.4$ MeV [1]
   c BE per nucleon = 7.5 MeV [1]
      BE = $7.5 \times 238 \approx 1800$ MeV [1]

5 mass of 8 protons = $8 \times 1.673 \times 10^{-27}$ kg [1]
   and mass of 8 neutrons = $8 \times 1.675 \times 10^{-27}$ kg
   BE = $[8 \times 1.673 \times 10^{-27} + 8 \times 1.675 \times 10^{-27} - 2.656 \times 10^{-26}] \times (3.00 \times 10^8)^2$
   BE = 2.016×10^{-11} J [1]
   BE = $\dfrac{2.016 \times 10^{-11}}{1.60 \times 10^{-19}} = 126$ MeV
   BE per nucleon = $\dfrac{126}{16} = 7.9$ MeV [1]

6 $\Delta m = 5.8 \times 10^{-3} \times 1.661 \times 10^{-27}$ kg [1]
   energy released = $\Delta m c^2 = 5.8 \times 10^{-3} \times 1.661 \times 10^{-27} \times (3.00 \times 10^8)^2$ [1]
   energy released = 8.67×10^{-13} J [1]
   total energy released = $8.67 \times 10^{-13} \times 6.02 \times 10^{23} \approx 5.2 \times 10^{11}$ J [1]

26.3/26.4

1 Energy is released in both fission and fusion reactions. [1]

2 a Z: left-hand side = 1 + 1 = 2 and right-hand side = 2 [1]
      A: left-hand side = 2 + 1 = 3 and right-hand side = 3 [1]
   b Z: left-hand side = 92 + 0 = 92 and right-hand side = 56 + 36 + 0 = 92 [1]
      A: left-hand side = 1 + 235 = 236 and right-hand side = 141 + 92 + 3 = 236 [1]

3 The nuclei are positive and therefore repel each other. [1]
   The (mean) kinetic energy of the nuclei is greater at higher temperatures and this means that nuclei can get close enough for the strong nuclear force to bind the nuclei together. [1]

4 energy = number of nuclei $\times 2.0 \times 10^{-11}$ [1]
   energy = $\dfrac{1.0}{0.235} \times 6.02 \times 10^{23} \times 2.0 \times 10^{-11}$ [1]
   energy = 5.1×10^{13} J [1]

5 The protons, positron, and neutrino are lone particles and therefore have no BE. [1]
   energy released = BE of $^2_1$H = $2 \times 1.0 = 2.0$ MeV [1]
   energy released = $2.0 \times 10^6 \times 1.60 \times 10^{-19} = 3.2 \times 10^{-13}$ J [1]

6 1.0 kg of hydrogen-1 has $\dfrac{1}{2} \times \dfrac{1.0}{0.001} \times 6.02 \times 10^{23} = 3.01 \times 10^{26}$ 'pairs' [1]
   energy = $3.01 \times 10^{26} \times 3.2 \times 10^{-13}$ [1]
   energy = 9.6×10^{13} J [1]

27.1/27.2

1 Simple scatter and Compton scattering. [1]
2 Photoelectric effect. [1]
3 $\dfrac{I}{I_0} = e^{-\mu x}$ [1]
   $\dfrac{I}{I_0} = e^{-0.21 \times 3.0} = 0.53$ [1]
   53% of the original intensity is transmitted through 3.0 cm of muscle. [1]

4 The maximum energy of the electron is 120 keV. [1]
   maximum energy of photon energy = 120 keV = $120 \times 10^3 \times 1.60 \times 10^{-19}$ [1]
   maximum energy of photon energy = 1.9×10^{-14} J [1]

Answers to summary questions

5 $eV = \dfrac{hc}{\lambda}$ [1]

$\lambda = \dfrac{hc}{eV} = \dfrac{6.63 \times 10^{-34} \times 3.00 \times 10^{8}}{1.60 \times 10^{-19} \times 200 \times 10^{3}}$ [1]

$\lambda = 6.2 \times 10^{-12}\,\text{m}$ [1]

6 For the photoelectric effect mechanism, $\mu \propto Z^{3}$. [1]

Therefore the attenuation of barium is $\left(\dfrac{56}{7}\right)^{3} \approx 510$ times greater than that of soft tissues. [1]

This makes it a much better absorber of X-rays than soft tissues. [1]

27.3/27.4/27.5

1 A three-dimensional image of the patient can be produced. [1]

2 X-rays are ionising or X-rays can damage cells. [1]

3 Lead absorbs gamma rays. [1]

Having narrow and long tubes allows only gamma photons from a specific direction to form the image (and hence produces a clearer image). [1]

4 PET scanning can be used. [1]

The FDG (or fluorine-18) will accumulate where rate of respiration is high and hence the **function** of the brain can be monitored. [1]

5 The time difference between the arrival times is $\dfrac{0.005}{3.0 \times 10^{8}} \approx 2 \times 10^{-11}\,\text{s}$. [1]

This means that the computer software has to process the information (or digital signals) very quickly. [1]

6 Tc-99m produces gamma photons that can pass through the patient. [1]

The gamma photons do not interact much with the atoms/cells of the patient. [1]

The short half-life of 6.0 hours means that the activity reduces to safe levels in a short period of time (e.g., the activity drops to about 6% of the initial activity after 1 day). [1]

27.6/27.7/27.8

1 A high-frequency alternating p.d. is applied to a piezoelectric material which makes it vibrate. [1]

The vibration of the material in air produces the ultrasound. [1]

2 In the time t the total distance travelled by the ultrasound is twice the thickness. [1]

3 $\Delta f \propto \cos\theta$; the change in frequency is zero when $\cos 90° = 0$. [1]

4 $\dfrac{I_r}{I_0} = \dfrac{(Z_2 - Z_1)^2}{(Z_2 + Z_1)^2}$ and $Z_2 = 2Z_1$ (or $Z_1 = 2Z_1$) [1]

$\dfrac{I_r}{I_0} = \left(\dfrac{1}{3}\right)^2$ [1]

$\dfrac{I_r}{I_0} = \dfrac{1}{9} \approx 0.11$ [1]

5 $\dfrac{I_r}{I_0} = \dfrac{(Z_{\text{air}} - Z_{\text{skin}})^2}{(Z_{\text{air}} + Z_{\text{skin}})^2}$ and $Z_{\text{air}} \ll Z_{\text{skin}}$ [1]

Therefore, the ratio $\dfrac{I_r}{I_0}$ is almost 1.0, with most of the ultrasound reflected at the air–skin boundary. [1]

6 $\Delta f = \dfrac{2fv\cos\theta}{c}$ [1]

$3200 = \dfrac{2 \times 18 \times 10^{6} \times v \times \cos 60°}{1600}$ [1]

$v = 0.28\,\text{m s}^{-1} = 28\,\text{cm s}^{-1}$ [1]